THE BIOLOGY OF
DESMIDS

BOTANICAL MONOGRAPHS

BOTANICAL MONOGRAPHS · VOLUME 16

THE BIOLOGY OF DESMIDS

ALAN J. BROOK

DSc, FRSE

Professor of Life Sciences
The University College at
Buckingham, England

UNIVERSITY OF CALIFORNIA PRESS

BERKELEY AND LOS ANGELES 1981

TO JOHN LUND

WITH RESPECT, ADMIRATION

AND GRATITUDE

UNIVERSITY OF
CALIFORNIA PRESS
Berkeley and Los Angeles

Library of Congress
Cataloging in Publication Data

Brook, Alan J.
  The biology of desmids.
  (Botanical monographs; v. 16)
  Bibliography: p.
  Includes index.
  1. Desmidiaceae. I. Title. II. Series.
QK569.D46B67 589'.47 80-26374

ISBN 0-520-04281-6

© 1981 by Blackwell Scientific
Publications

Printed in Great Britain

CONTENTS

PREFACE

Like the blue-green algae and particularly the diatoms with which they are often confused (and which have been the subjects of earlier monographs in this series), the desmids are a group of algae which have fascinated microscopists for over 150 years. In consequence, there are some 8000 references relating to these beautiful organisms scattered throughout the botanical and limnological literature. The majority of these papers relate to their morphology and taxonomy, and since 1848 when the Rev. John Ralfs produced the first book for their identification, British Desmidieae, a succession of keys have been published in Europe, Russia, the United States and Japan. As far as I am aware, however, no one has produced a biology of desmids, and so this volume has been prepared in an attempt to fill that gap.

The attempt is a very personal one, and no doubt some readers may feel that there are omissions, inaccuracies, or a lack of emphasis (or too much) on particular topics. I take the entire responsibility for all the book's possible shortcomings; but in saying that, what I have attempted to do is not merely summarize what is currently known about desmids but also to point to those aspects of desmid biology which could prove to be exciting and fruitful fields of research. What I have written will, I hope, stimulate not only students of the algae, but also those in other disciplines, to look at and use the potential contained in these beautiful organisms in cytological, genetical, ecological and cell physiological research. Regretably students of desmids are all too few.

A.J.B.

Shalstone, Buckingham

ACKNOWLEDGMENTS

I wish to acknowledge a very special debt of gratitude to my secretary at the University College at Buckingham, Miss Julie Cakebread, who has patiently typed and retyped the manuscript speedily and efficiently. I would like to thank Mrs Ruth Clarke and Mrs Irene Gray of the print room at Buckingham for their unstinting cooperation in copying papers, manuscripts and illustrations. Much valuable library help has been provided by Mr Ian Pettman of the Freshwater Biological Association; also by Mr Richard Newell of Buckingham, and several members of the staff of Reading University Library.

Thanks are due to Mr Alan Jenkins of the Department of Zoology at the University of Reading for hours spent in producing prints of photomicrographs from often less than adequate negatives; to Dr Peter Brandham of the Herbarium, the Royal Botanic Gardens, Kew, for allowing me to use many of his published and unpublished illustrations of cytological events in desmids; to Dr Hilda Canter-Lund of the Freshwater Biological Association for all the time taken in helping me to select from her large collection of superb photomicrographs of desmids, for making such excellent prints from them and allowing me to use them.

I am indebted to Dr Tim Lack of the Water Research Centre, Medenham, for unpublished information about a bloom of desmids at Siblybae Reservoir, Cornwall, and about grazing experiments carried out in his laboratory. Dr Colin Reynolds has been most helpful in providing details of the results of experiments carried out in Lund Tubes in Blelham Tarn. To my daughter Erica, of Chur, Switzerland, I am truly grateful for much assistance with the translation of several papers in German, as I am also to Dr David Hunniford of the Language Department at the University College at Buckingham. Finally I am most grateful to the staff of the Production Department of Blackwell Scientific Publications for their patience and understanding of a not always very well organized author.

CHAPTER 1
MORPHOLOGY AND PHYLOGENETIC
RELATIONSHIPS

INTRODUCTION

Desmids, which are exclusively freshwater algae, are so named because the first of these green microorganisms to be described were thought to be pairs of cells joined together, and the Greek word 'desmos' means a bond or chain. However, subsequent investigations have indicated that what were first considered as cell pairs are in fact two half cells (semicells) and it is now abundantly clear that by far the greatest number of desmid species are unicellular. Unfortunately, the initial misleading implication that desmids characteristically form chains remains.

The unicellular desmids attracted the attention of early microscopists (Ralfs 1848) for these forms, in particular, exhibit great diversity in their external morphology and show a remarkably complex cell symmetry, making them organisms of great natural beauty and hence aesthetic appeal. As a consequence of their diversity (and undoubtedly their beauty), more than 6000 species have been described from freshwaters in all parts of the world. Interest in external morphology later extended to studies of their internal features, their possible origins and phylogenetic relationships, their reproduction, both asexual and sexual and also their genetics. Aspects of the development and morphogenesis of elaborate desmid cells and investigations of the submicroscopic structure of both the internal structure and elaborately sculptured walls have been conducted with transmittance and scanning electron microscopy, respectively. Because of their almost ubiquitous occurrence in freshwaters and the fact that quite distinctive assemblages of species occur in particular habitats, considerable attention has been focussed on desmid ecology. It is these wide-ranging topics which make desmids objects of considerable scientific interest and about which the following chapters deal.

1

DESMIDS IN RELATION TO OTHER PLANTS AND ALGAE

Algae are amongst the simplest of plants. Normally they possess chlorophyll in all their cells and most of them grow in marine and freshwater environments, though a few are terrestrial. In addition to the great diversity of unicellular algae, which may be solitary or colonial, there are numerous multicellular forms often of macroscopic size, such as the giant kelps. All can be distinguished from other photosynthetic cryptogamic plants by their unicellular organs of reproduction, their relatively simple organisation and, in particular, by their lack of true leaves, stems or roots.

The term algae, Latin for seaweeds, does not refer to any natural assemblage of plants, rather it is a term of convenience embracing as many as ten distinct and unrelated divisions or phyla of the Plant Kingdom; for example the Cyanophyta, Rhodophyta, Chrysophyta, Phaeophyta, Pyrrophyta, Euglenophyta and Chlorophyta, etc. The separation of these phyla is essentially based on their biochemistry, and depends largely on the type of photosynthetic pigments present and the nature of the stored food reserves. The Cyanophyta differ from the others, however, in that they possess no true nucleus and this feature, and others, indicate their close relationship with the bacteria. The Chlorophyta, to which phylum the desmids belong, possess grass-green chloroplasts containing pigments essentially the same as those of all higher green plants (see, however, Chapter 3, p. 43), and starch is lodged in the chloroplasts as a reserve product of photosynthesis.

Phyla of the Plant Kingdom are further divided into classes and, in recent years, there has been some acceptance of the taxonomic system whereby the Chlorophyta is split into four classes. These are:
Class Chlorophyceae
Class Bryopsidophyceae
Class Zygnemaphyceae
Class Oedogoniophyceae.
The Class Zygnemaphyceae, which until fairly recently was termed the Conjugatophyceae (Round 1963, Bourrelly 1966), was proposed by Round (1971) in his extensive revision of the taxonomy of the Chlorophyta, on the grounds that the names of classes should be derived from a generic name. It is typified and hence distinguished from other classes within the Phylum Chlorophyta by:
1 the absence at any stage in the life cycle of flagella which in other classes confer motility, especially during reproduction;
2 reproduction by the conjugation of non-flagellated, amoeboid gametes, for which reason the name Conjugatophyceae was previously applied to the class.

The initial attraction even now for many students who study desmids is their morphology, for in addition to some being the largest of unicellular plants, they show, as a group, a greater range and complexity of form than any unicellular organisms. Some are almost spherical and as small as 10 μm, or less, in diameter, while rod-like forms may attain lengths of over 1000 μm (1 mm) and hence are visible to the naked eye.

Collections of algae, especially from the weedy margins of lakes and ponds, or ditches and bogs, particularly in base-poor, granitic regions, can contain an almost bewildering array of desmids and it is not uncommon for over 100 different species to be identified from a single 'good' habitat. Because desmid cells can, under certain conditions, show considerable morphological plasticity, a given species from such collections may be present in a range of 'growth forms'. Many of these have been given taxonomic status especially by the earlier investigators and so desmid taxonomy tends now to be cluttered by a considerable number of doubtful 'species' or 'varieties'. Many of these, when populations are examined, rather than merely randomly chosen individuals, will undoubtedly be shown to be growth forms. Also because of their plasticity even the demarcation of genera is not always clear. Indeed much remains to be accomplished in the revision of desmid taxonomy (see pp. 152 and 180).

Before attempting to survey the range of form of the desmids and exploring their possible phylogenetic relationships it is appropriate to enquire into the relationship of the class Zygnemaphyceae (of which the desmids constitute two important orders) with other classes of the phylum Chlorophyta. The Zygnemaphyceae are clearly unique in their mode of cell division, their lack of zoospores or motile gametes, and their sexual reproduction by a process of conjugation. Like other groups of green algae such as the stoneworts (Charophyta) and Oedogoniophyceae they appear to be surviving relics of uncertain ancestry, or as Prescott (1968) suggests, 'they are one of the ultimate twigs of an evolutionary tree whose lower branches disappeared long ago'.

Pickett-Heaps and Marchant (1972) emphasize that many of the criteria upon which numerous currently accepted phylogenies have been based are of uncertain value and the consequent evolutionary schemes have certain unaccounted for discrepancies. Their researches, which involve comparisons of the morphology of the mitotic and cytokinetic apparatus, suggest that a fundamental distinction can be made between those groups that possess, or lack, the phycoplast system of microtubules involved in cytokinesis. They suggest that the Zygnemaphyceae, and some of the Chaetophorales and Charophyta, possess a persistent mitotic spindle and they intimate how the phragmoplast, characteristic of cytokinesis in higher plants, may have evolved from this.

The diagram below (Fig. 1) is a modification of their phylogenetic scheme which shows the distinction between phycoplast-containing green algae and those possessing a persistent interzonal spindle which are considered to be related to the progenitors of the higher plants. Other interesting views on the origins of the major groups of green algae and land plants, and the evolution of cell walls, from presumed scaly phytoflagellates, have been offered by Stewart and Mattox (1975, 1978).

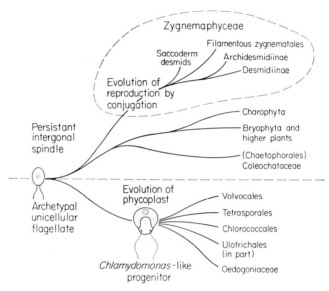

Fig. 1. Diagrammatic representation of the evolution of the Zygnemaphyceae, and their relationship to other green algae (after Pickett-Heaps and Marchant 1972).

Although we shall probably never know how, why or when the Zygnemaphyceae arose, it is not a great leap from an assumed archetypal, uniflagellate ancestor to the loss of the flagellate condition (which has happened in other groups of green algae) and the adoption of a conjugate mode of sexual reproduction.

THE SACCODERM DESMIDS

(Order I: ZYGNEMATALES) (Family: MESOTAENIACEAE)

Closest to what must be assumed to be the conjugate archetype are representatives of the Saccoderm desmids of the family Mesotaeniaceae. *Mesotaenium*, the type genus, is typified by its short cylindrical cells whose smooth cell walls are without pores; nor do they have a median suture line. Each cell contains an apparently simple chloroplast in which there are one or two pyrenoids (see however Dorscheid and Wartenberg 1966 and p. 66). *Mesotaenium*-like cells which form short filaments of 2–16 cells in length constitute the monospecific genus *Ancylonema*. *A. nordenskioldii* is found on mountain snows almost throughout the world.

Cells also without pores or a median suture but which are much longer in relation to their width than *Mesotaenium* and usually slightly curved, constitute the genus *Roya*. Each *Roya* cell contains a single ribbon-like chloroplast but it is notched in the middle where the nucleus occurs. Depending on the species, each chloroplast contains from 4 to 12 pyrenoids.

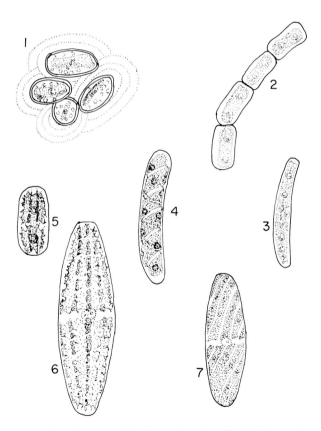

Fig. 2. Representatives of genera of the Saccoderm desmids: 1 *Mesotaenium macrococcum* Roy and Biss.; 2 *Ancylonema nordenskioldii* Bergr.; 3 *Roya obtusa* (Bréb.) W. & G.S. West; 4 *Spirotaenia condensata* Bréb.; 5 *Cylindrocystis brebissonii* Menegh.; 6 *Netrium digitus* (Ehr.) Itz. & Rothe.; 7 *Spirotaenia* (?) *endospora* Bréb. (see p. 48).

Cells with a much more complex chloroplast, reminiscent of *Spirogyra*, typify the genus *Spirotaenia*. The elongate cells, which may be cylindrical or slightly fusiform, contain in some species a parietal, spirally wound chloroplast embedded within which is a series of pyrenoids; in others there is a large central (axile) chloroplast with helically arranged projections (see also p. 45), and thus should possibly be assigned to a new, separate genus.

In *Mesotaenium*, *Ancylonema* and *Spirotaenia* the chloroplast extends from end to end of the cell without interruption, while in *Roya* it is notched in the middle to accommodate the nucleus. In the two remaining genera of the Mesotaeniaceae, *Cylindrocystis* and *Netrium*, each cell contains two chloroplasts, one in each end of the cell and separated by a centrally placed nucleus. This presumably more advanced feature, is a characteristic of the more highly evolved placoderm desmids. In both genera, the chloroplasts are axile and star-shaped in cross-section, but the indentations of the margins are much less pronounced in *Cylindrocystis* than in *Netrium*; the cells of the former are always cylindrical with rounded ends, whereas in the latter they are mostly truncate-elliptical (Fig. 2.1, 1–7; see also Fig. 22 and p. 44).

THE PLACODERM DESMIDS

(Order: DESMIDIALES)

The principal features that distinguish the placoderm from the saccoderm desmids are that each cell of the former consists of two parts, with a much more complex two-layered wall, perforated by a system of pores. Details of these features and of the unique mode of cell division of placoderm desmids will be outlined later (Chapter 5).

It is quite apparent that there is a very considerable evolutionary gap between the saccoderm and placoderm desmids and, except for the fact that there is a chloroplast in each end of the cell in *Cylindrocystis* and *Netrium*, and some similarities in the germination of zygospores in the latter genus, there is no evidence to suggest how the very distinctive features of the placoderms might have arisen from the archetypic ancestor. The least advanced of the placoderms belong to the

	Suborder:	Archidesmidiinae
with three	Families:	Gonatozygaceae
		Peniaceae
		Closteriaceae

All other placoderms are included in the:

	Suborder:	Desmidiinae
with but one	Family:	Desmidiaceae

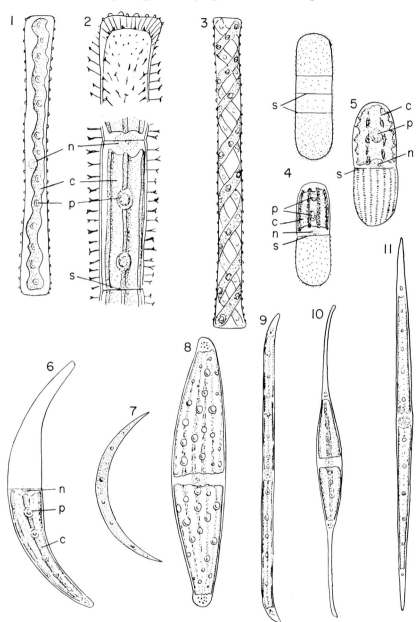

Fig. 3. Placoderm desmids of the suborder Archidesmidiinae: 1 *Gonatozygon monotaenium* De Bary; 2 *G. chadefaudii* Bourrelly; 3 *Genicularia spirotaenia* De Bary; 4 *Penium cylindrus* (Ehr). Bréb.; 5 *P. silvae-nigrae* Rabenh. (after Bourrelly); 6 *Closterium leibleinii* Kütz; 7 *C. exiguum* W. & G. S. West; 8 *C. lunula* Ehr.; 9 *C. gracile* Bréb.; 10 *C. rostratum* Ehr.; 11 *C. pronum* Bréb. (c=chloroplast; n=nucleus; p= pyrenoid; s=suture line.)

SUBORDER: ARCHDESMIDIINAE

Until a few years ago the genera *Gonatozygon* and *Genicularia* were classified as saccoderm desmids within the Family Mesotaeniaceae because of their presumed simple wall structure and mode of division. Both genera are characterized by their very long, cylindrical, or fusiform cells, the walls of which are often granulate, or in some species of *Gonatozygon* covered by fine spines (Fig. 3: 1, 2). After division the cells sometimes adhere together, so that short filaments are formed. The two genera are separated largely by the form of their chloroplasts. In *Gonatozygon* it is a simple axial ribbon which may be straight or slightly undulating and containing a series of pyrenoids down its length; *Genicularia* differs in that there are 2 or 3 parietally placed ribbon-shaped, spirally wound chloroplasts also containing a series of pyrenoids (Fig. 3: 3). Detailed EM studies by Mix (1972) comparing the wall structure of various genera and species of Mesotaeniaceae such as *Cylindrocystis* spp., *Roya* spp., *Netrium* spp. and *Mesotaenium caldariorum* with three species of *Gonatozygon*, as well as *Mougeotia* and *Zygnema* spp. (members of the Family Zygnemataceae), has indicated that *Gonatozygon* can no longer be classified as a saccoderm desmid. All saccoderms, along with *Mougeotia* and *Zygnema*, possess walls without pores or sculpture, whereas *Gonatozygon* (and it is presumed *Genicularia*) has a very delicate pore system and sculptured walls. Since these are unquestionably placoderm characteristics, these two genera are either to be included in the family Closteriaceae along with *Penium* and *Closterium* (Bourrelly 1966), or even better, as suggested by Mix (1972), placed in a separate family Gonatozygaceae (see p. 96). Of considerable interest and significance is the description of a new species of *Gonatozygon* by Bourrelly (1977), *G. chadefaudii*, which shows transverse suture lines on its cell walls, a feature not seen in other species of this genus (Fig. 3: 2).

With respect to cell shape, the genus *Penium* represents the simplest of the placoderm desmids. These typically are cylindrical cells with rounded ends and with walls perforated by a system of simple pores. Each bears a median suture line demarcating each cell into two semicells. Cells increase in length at specific zones of elongation with the result that several suture lines can be seen. The walls may be ornamented with pores arranged in distinct longitudinal rows or they may be irregularly arranged (Fig. 3: 4, 5).

Clearly related to *Penium* is the large and morphologically diverse genus *Closterium* (Fig. 3: 6–11) whose cells are most frequently lunate with tapering or pointed poles, though a few species approach *Penium* in that they are straight and only the ends are curved, and then very slightly. At least one median suture line can always be seen, and in some species, because of clear zones of elongation, several suture lines may be apparent in one cell (see p. 104 and Fig. 45). The cell walls which, though usually colourless, may be yellow-brown due to incrustation by iron salts, often appear smooth though they are always perforated by pores. The walls of some species bear well defined longitudinal striations (Fig. 37) (p. 83). On the basis of some fairly fundamental morphological differences it has been proposed by Bourrelly (1966, p. 404) that the genus *Closterium* be divided into 6 sections (see also Růžička 1975a and b, 1976, 1977).

SUBORDER: DESMIDIINAE

All other placoderm desmids, and by far the majority of species, are those which exhibit a more or less well defined median constriction, the isthmus, so that each cell is clearly composed of two semicells. The adjoining semicells, as will be explained later (p. 108), are separate entities of different ages. They are nevertheless closely joined by overlapping, hooked walls and there is never any zone of elongation between the two. All desmids with these characteristics are embraced in the Family Desmidiaceae.

The simplest of the constricted placoderms are encountered in the genus *Pleurotaenium* whose cells consist of straight, baton-like cylinders with truncated ends (Fig. 4: 1, 2). In some species there is a slight swelling just above the isthmus, while in others the entire lateral margins show regular undulations. The cell apices may be smooth or encircled by a crown of tubercules or small warts. The cell walls, though always porous, may appear smooth or be elaborately decorated, depending on the species, with regularly arranged spines or scrobiculations. *Docidium* differs from *Pleurotaenium* in that around the median constriction of the elongated cylindrical cells there is on each adjoining semicell, a crown of warts (Fig. 4: 3).

The cells of *Triploceras* are always solitary elongated cylinders with only a very slightly constricted isthmus. They are characterised by the fact that the two poles are flattened and extend into two short lobes each of which terminates in two or three spines. Between the lobes is a small projecting tumour bearing one or two spines. The entire cell wall is ornamented by rows of regularly arranged simple spines or verrucae (Fig. 4: 4). In *Triplastrum*, of which there are only two rarely found species, the cells (as in *Triploceras*) are solitary, cylindrical and with only a slightly obvious isthmus. However, it differs in that the apices are slightly dilated and divided into three short divergent lobes terminating in from one to four spines (Fig. 4: 5).

Another genus with obvious affinities with *Pleurotaenium* is the tropical *Ichthyocerus* (fish-tail), so named because the apices of its elongate cylin-

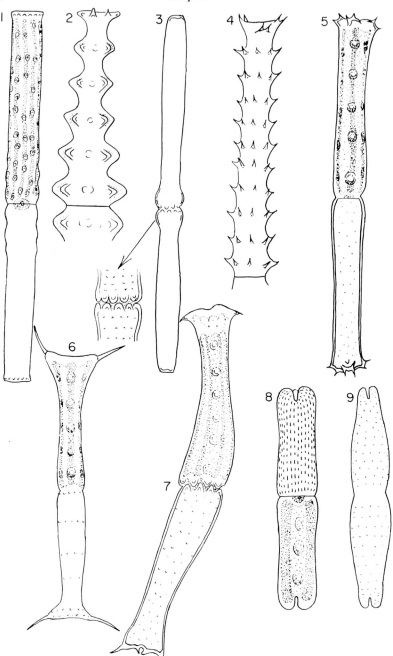

Fig. 4. Placoderm desmids of the suborder Desmidiinae: 1 *Pleurotaenium cylindricum* var. *stuhlmanni* (Hieron.) Krieg. (after Bourrelly); 2 *P. nodosum* (Bail.) Lund.; 3 *Docidium baculum* Bréb. (after Bourrelly); 4 *Triploceras gracile* Bailey (after A. M. Scott); 5 *Triplastrum simplex* (Allorge) Iyengar & Raman; 6 *Ichthyocerus longispinus* (Borge) Krieger; 7 *Ichthyodontum sachlanii* Scott and Prescott; 8 *Tetmemorus brebissonii* (Menegh.) Ralfs; 9 *T. granulatus* (Bréb.) Ralfs.

drical cells are slightly enlarged to form two lobes like the tail of a fish. Each lobe ends in a single spine (Fig. 4: 6).

In 1956 a desmid with characters found in both *Ichthyocerus* and *Docidium* was described from Java and named *Ichthyodontum sachlanii* (Scott and Prescott 1956). Thus, like *Docidium* it has scrobiculations around its isthmal region but which fit one into the other; also it has the fish-tail-like apices of *Ichthyocerus*. One quite distinctive feature of this new desmid is that the cells are curved, or in some cases even sigmoid; also adjacent semicells are asymmetric with respect to one another (Fig. 4: 7) (see p. 13).

Another *Pleurotaenium*-like genus, *Tetmemorus*, is characterised by semicell apices which show a pronounced and often deep median cleavage. The cell walls of members of this genus are scrobiculate and the scrobiculations tend to be arranged in longitudinal rows (Fig. 4: 8, 9).

A genus which would seem to occupy a central position in phylogenetic schemes, in that it appears to provide a link between the supposedly primitive *Penium* and the more advanced genus *Cosmarium*, is *Actinotaenium*. This genus includes some 40 species and has been the subject of a detailed taxonomic revision by Teiling (1954). (See, however, Růžička and Pouzar 1978.) It is typified by cells which are always circular in transverse section and are cylindrical to fusiform with rounded or slightly flattened poles, but with only a slight median constriction separating the semicells. The walls are without ornamentation but are porous except in the isthmal region; unlike those of *Penium*, which superficially they may closely resemble, there is no evidence of any elongation after cell division, so that there is always only a single isthmal suture line. Also, although pores are clearly evident in the walls, they are never as clearly arranged, as in *Penium*, in parallel longitudinal rows (Fig. 5: 1–3).

Cosmarium is the genus which typifies a desmid for most people. The cells are usually solitary having a well-marked median constriction and semicells which are entire in outline, without lobes of any sort (Fig. 5: 4–11). In the majority of species, of which there are many thousands, the apical view is elliptical in outline though some can be seen to be circular. Although the punctate walls are in many cases elaborately ornamented by scrobiculations and granules, they do not possess the prominent spines which characterise the related genera *Arthrodesmus* and *Xanthidium*. The characters of taxonomic importance in the separation of the thousands of species of *Cosmarium* are cell shape and wall ornamentation. It should be emphasised that this large genus is a purely artificial one and morphological series indicate its close affinities with the genera *Arthrodesmus*, *Xanthidium*, *Euastrum*, and especially *Staurastrum*.

As emphasized by Bourrelly (1966) it is not easy, because of their artificial nature, to find criteria which permit a logical systematics scheme

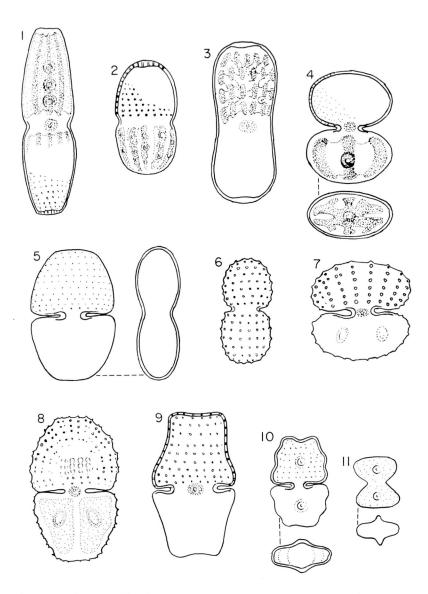

Fig. 5. Placoderm desmids of the suborder Desmidiinae: 1 *Actinotaenium elongatum* (Racib.) Teil.; 2 *A. capax* var. *minus* (Nordst.) Teil.; 3 *A. diplosporum* (Lund.) Teil. (after Teiling); 4 *Cosmarium contractum;* 5 *C. pyramidatum* Bréb.; 6 *C. amoenum* Bréb.; 7 *C. reniforme* (Ralfs) Arch.; 8 *C. formosulum;* Hoff.; 9 *C. decedens* (Reinsch) Racib.; 10 *C. venustum* var. *excavatum* (Bréb.) Arch.; 11 *C. staurastroides* Eichl. et Gütw. (Eichl. et Gütw.) West.

for the huge genus *Cosmarium*. However, following the suggestions of Cedergren (1933) and Hirano (1955, 1956, 1957 and 1959), Bourrelly divides it into 3 sub-genera:

1 Sub-genus *Dysphinctium* with cells circular or sub-circular in apical view.

2 Sub-genus *Cosmarium* with cells elliptical in apical view.

3 Sub-genus *Nothocosmarium* with cells in apical view semi-circular, semi-elliptical, reniform or rarely triangular.

The genus *Euastrum* must be assumed to have evolved from *Cosmarium*-like cells of the sub-genus *Cosmarium*, for its members are all elliptical in apical view. It is distinguished, however, by the fact that each semicell apex is cleft by a more or less well defined incision. The front view of the semicells is basically pyramidate, and although a few species have smooth cell walls, the majority are elaborately ornamented by wart-like granules or protruberances, which are especially pronounced in the centre of the basal region of each semicell (Fig. 6: 1–3). It can be difficult to distinguish some species of *Euastrum* from *Cosmarium* when the apical notches or clefts are very slight (Fig. 5: 10). On the other hand some species of *Euastrum* whose outlines are elaborately lobed approach some of the less elaborate species of the genus *Micrasterias*, and 'vice versa', as for example, *M. euastriellopsis* (Bharati 1965) (Fig. 6: 4).

The genus *Micrasterias* is typified mostly by very large and spectacularly shaped cells which are solitary (except in the species *M. foliacea* which occur in chain-like filaments), and appear very markedly flattened in apical view. They are circular to quadrangular in outline in front view, and although in some species the apices of the semicells may be entire (thus unlike *Euastrum*) (Fig. 6: 6), the margins in most species are cut by several deep incisions, giving a distinctly lobed outline (Fig. 6: 7). There is always a very pronounced median constriction so that adjoining cells are joined by a comparatively narrow isthmus. The walls of some species are smooth, though porous, but in many ways they are elaborately ornamented by regularly arranged granules, scrobiculations or spines.

Two species of *Micrasterias*-like desmids have been described in the past twenty years from Central Africa and indeed one was originally named *Micrasterias incredibilis*. However, they differ in that they are asymmetric desmids (see Chapter 2) in which one semicell differs consistently from that which it adjoins. Thus, one has an apical lobe bearing simple spines (Fig. 7: 1) whereas the other has an apical lobe with numerous sub-lobes bearing simple or bifurcate spines (Fig. 8: 1). The two species are so distinctive, as well as morphogenetically fascinating, that one has been placed in a separate genus *Allorgeia* (Gauthier-Lièvre 1958), the other in *Prescottiella* (Bicudo 1973).

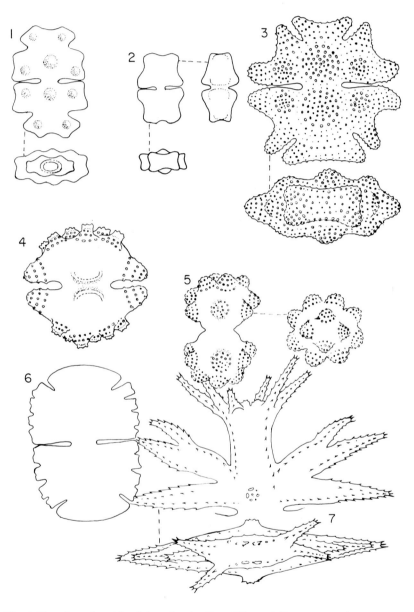

Fig. 6. Placoderm desmids of the suborder Desmidiinae: 1 *Euastrum pectinatum* var. *inevolutum* W. & G. S. West; 2 *E. sublobatum* var. *subdissimile* W. & G. S. West; 3 *E. verrucosum* Ehr.; 4 *Micrasterias euastrellopsis* Bharati (after Bharati); 5 *Euastridium verrusosum* Carter (after Carter); 6 *Micrasterias crenata* Bréb.; 7 *M. mahabuleshwarensis* var. *wallichii*.

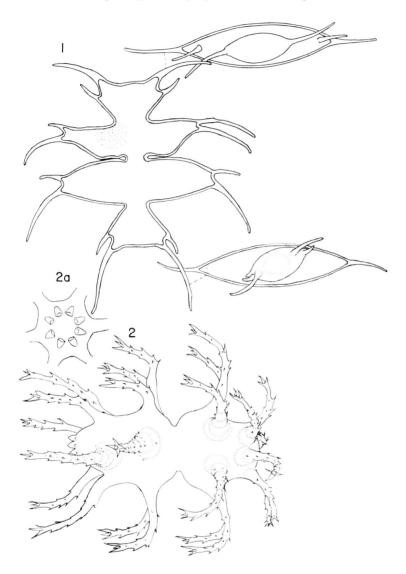

Fig. 7. Placoderm desmids of the Suborder Desmidiinae: 1 *Prescottiella sudanensis* (Grönbl., Prowse & Scott) Bicudo; 2 *Amscottia mira* (Grönbl.) Grönbl.

Another form of cell elaboration seen in *Cosmarium*-like desmids is exhibited by members of the genus *Xanthidium*. Like *Cosmarium*, from which they are clearly derived, they are elliptical in apical view but each side of the ellipse bears a marked swelling which is often ornamented with large pores or is scrobiculate. In front view the semicells are most commonly hexagonal or octagonal, a fact emphasised in that each angle of these polygons bears a pronounced spine (Fig. 8: 3, 4). When viewed apically it can be seen that the spines are arranged in two series, one on each side of the semicell. In a few species, as for example in the large and fairly common *X. armatum*, the spines are forked (Fig. 38: 4; see Chapter 4).

A new genus *Bourrelloydesmus* has been proposed by Compère (1976) for a *Xanthidium*-like desmid named originally *Arthrodesmus heimii* (the problems of desmid taxonomy!!) and found in several parts of Africa. Whether it is valid to create such a genus is open to question since the desmid concerned would seem to differ from *Xanthidium* only in the possession of a single spine on each side above the isthmus of each semicell (Fig. 8: 5).

Linking the genera *Cosmarium* and *Xanthidium* is the genus *Spinocosmarium* with only two species, both of which have been found only in North America. As in many species of *Xanthidium* the semicells are hexagonal, but above the isthmus there is a single strongly developed spine which is frequently bifurcate (Fig. 8: 2). The apical margins, as in *Xanthidium*, bear pairs of spines which in vertical view are seen to be diagonally disposed. The cell walls are ornamented with more or less parallel series of veruccae, but there is no lateral swelling as in *Xanthidium*.

Although the vast majority of *Cosmarium* species are solitary; rarely forming filaments and then of only a few cells in length, there appears to have been an evolutionary exploration of the filamentous and the colonial habit. Thus in *Cosmocladium* small, elliptical and mostly unornamented *Cosmarium*-like cells are held together by gelatinous strands secreted through two series of pores, one each side of the isthmus, but restricted to one face of the cell (Gerrath 1970) (see p. 97). In addition, a gelatinous sheath may envelop the entire colony (Fig. 9: 3, 4).

An even more remarkable colonial *Cosmarium*-like species occurs in swiftly running calcareous streams, firmly attached to rocks. The cells of this desmid, *Oocardium stratum*, are united into sessile colonies consisting of broad, dichotomously branched, lime-encrusted, hollow gelatinous tubes, a single cell occurring at the end of each (Fig. 9: 1, 2). (See also Rieth 1970.)

The filamentous colonies of *Spondylosium* consist of *Cosmarium*-like cells in which the apices of adjoining cells are connected to one another by mucilage pads which develop between their contiguous faces (Fig. 9: 5, 6). The similar genus *Sphaerozosma* differs, however, in that the adjoining cells are held together by short apical processes that interlock with one another. In some species these processes may be reduced to mere granules (Fig. 10: 1) (Kirk, Postek and Cox 1976).

Members of the filamentous genus *Onychonema* are typified by the occurrence on each semicell apex of two asymmetrically disposed, quite

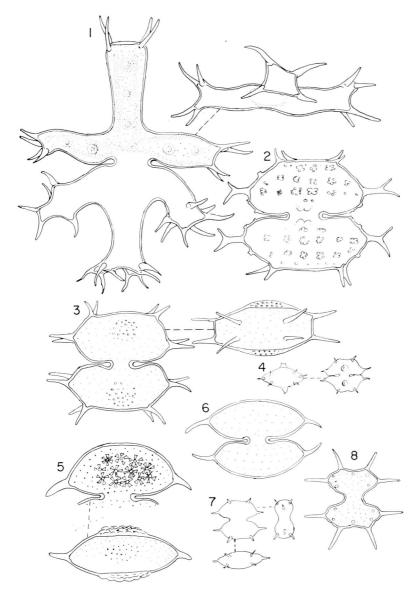

Fig. 8. Placoderm desmids of the suborder Desmidiinae: 1 *Allorgeia incredibilis* (Grönbl., Prowse & Scott) Thomass. (after A. M. Scott); 2 *Spinocosmarium quadridens* fo. *forficulata* Presott and Scott; 3 *Xanthidium antilopaeum* (Bréb.) Kütz.; 4 *X. concinnum* Arch. (after Bourrelly); 5 *Bourrellyodesmus heimii* (Bourr.) Compère (after Bourrelly); 6 *Staurodesmus convergens* (Ehr.) Florin; 7 *Arthrodesmus tenuissimum* Arch. (after W. & G. S. West); 8 *A. octocornis* Ehr.

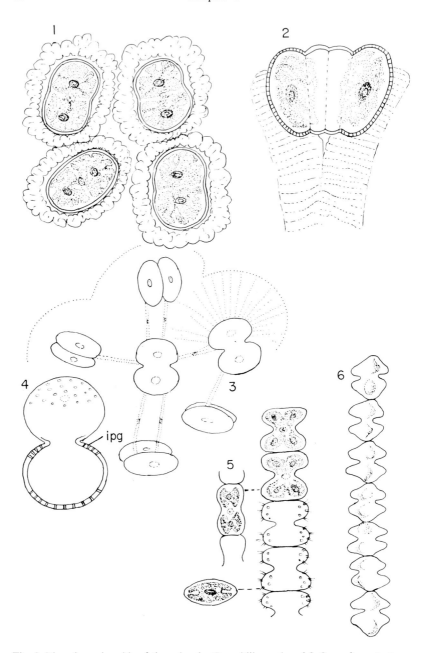

Fig. 9. Placoderm desmids of the suborder Desmidiinae: 1 and 2 *Oocardium stratum* Näg; 3 and 4 *Cosmocladium saxonicum* De Bary; 5 *Spondylosium planum* (Wolle) W. & G. S. West; 6 *S. moniliforme* Lund.

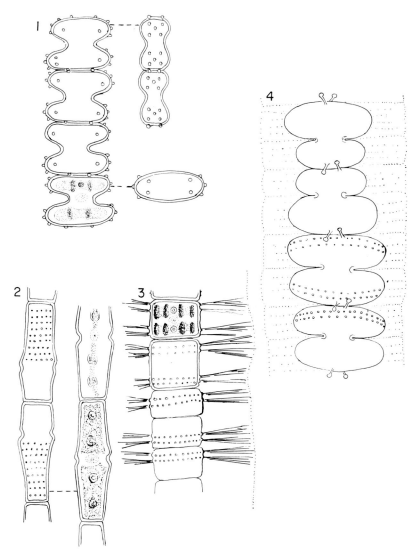

Fig. 10. Placoderm desmids of the suborder Desmidiinae: 1 *Sphaerozosma granulata* Roy and Biss.; 2 *Groenbladia neglecta* (Racib.) Teil. (after Bourrelly); 3 *Hyalotheca mucosa* (Dillw.) E.; 4 *Onychonema filiformis* (Ehr.) Roy and Biss.

long capitate processes which overlap the adjoining cell and firmly unite one cell to the next (Fig. 10: 4). (See also Bourrelly 1964.)

One of the commonest filamentous desmids, some species of which are planktonic, is *Hyalotheca* (e.g. *H. dissiliens*). Because the median constriction is so slight as to be barely perceptible in such species as *H. mucosa*, it is at first difficult to realize that *Hyalotheca* is a desmid. The cells are short cylinders with flat apices, by which adjacent cells are firmly joined. There is no ornamentation on the walls except for delicate transverse ridges which occur just below the cell apices (Fig. 10: 3).

As pointed out above, one of the shapes which may be seen when *Cosmarium*-like cells are viewed apically, is a triangular outline (cf. subgenus *Nothocosmarium*). If such forms are consistently triangular they are placed in the genus *Staurastrum* (see below). However, like these forms of *Cosmarium*, a few species of *Euastrum*, *Xanthidium* and *Micrasterias*, all of which typically can be seen to be elongate-elliptical, or markedly flattened in apical view, may occur as triangular forms (Fig 11: 1). These are essentially identical in front view with the species from which they are obviously derived. Their significance is discussed in Chapter 2. As pointed out by Fritsch (1952) with respect to *Cosmarium*, the nomenclature of the forms which may be either elliptical or triangular in apical view is quite illogical, since for many of the simpler trigonal species of *Staurastrum* with 4, 5 or more angled apical views, it is possible to find almost exact replicas among species of *Cosmarium*. It was because of this illogicallity that Fritsch (loc. cit.) suggested that it might be appropriate to group together all the *Cosmarium*-like forms, which in apical view are 3 or more sided and the angles of which are not produced as processes, under one common generic heading (Fig. 11: 1, 2). He proposed the name *Cosmostaurastrum*, but his very reasonable proposal has not yet been adopted by desmid taxonomists (see however, p. 22 and Palamar-Mordvintseva 1976).

In apical view many triangular desmids appear modified so that each of the angles is more or less extended or drawn out into hollow processes (Fig. 11: 3 and Fig. 12: 1–5). This extension of the angles, which is also apparent when the cells are examined in front or side view, is a characteristic of the majority of species of *Staurastrum*, and indeed the type species of this genus (Meyen 1828) was a desmid of this shape (see however Brook 1959c) (Fig. 12: 3). Because of the usually marked extension of the polar angles of the semicells, the appearance of *Staurastrum* species in both apical and front view is clearly different from those triangular species ascribed to *Cosmarium* (or Fritsch's *Cosmostaurastrum*), although, as with nearly all genera of desmids, there is no clear cut distinction between the two genera (Fig. 11: 3).

The genus *Staurastrum* exhibits the greatest range of morphologies of all desmid genera and is a very artificial taxon. The majority of species have cells that are radially symmetrical (see Chapter 2) and usually triangular in apical view, there are some however, whose cells are markedly bilaterally compressed and others that are quadrangular, pentangular, or

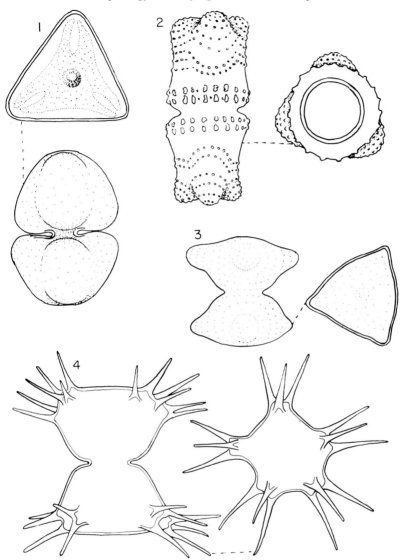

Fig. 11. Placoderm desmids of the suborder Desmidiinae: 1 *Staurastrum orbiculare* var. *ralfsii* W. & G. S. West; 2 *S.* (*Cylindriastrum*) *pileolatum* Bréb.; 3 *S. subpygmaeum* West; 4 *S.* (*Raphidiastrum*) *brasiliense* var. *lundellii*. W. & G. S. West.

hexangular to 9-angular. Most have cells that are deeply constricted so that there is a well defined isthmus. The walls, always with a pore system which may or may not be clearly visible with the light microscope, may be smooth or ornamented with granules, denticulations, verrucae or spines which are always arranged in a consistent symmetrical pattern of taxonomic significance. The pores often extrude mucilage as plugs, or it may envelop the entire cell. The front view of the semicells may be elliptical, semicircular, cyathiform, triangular, quadrangular or polygonal in outline and especially in those species which occur in the plankton, the upper angles are extended into long hollow processes. These are variously ornamented, and terminate in truncate ends bearing usually 3 or 4 short divergent spines. In a few species, again frequently found in the plankton, two transverse whorls of processes are borne on each semicell. Although the triradiate or trigonal form is most commonly encountered in *Staurastrum*, some species typically vary with respect to this character. Thus forms which are 2-, 3- or more radiate may all occur in a given population. This phenomenon will be discussed in detail in Chapter 2. The most acceptable classification of the genus is that of Hirano (1955–60).

The production of spines rather than hollow processes at the angles of smooth-walled, bilaterally compressed, *Cosmarium*-like cells (Fig. 8: 6–8 and Fig. 13: 2) typifies the genus *Arthrodesmus* which, as stated by Bourrelly (1966), requires a substantial, critical revision. Until 1948 similar triangular forms were quite illogically ascribed to the genus *Staurastrum*. In that year Teiling (1948) proposed the establishment of the genus *Staurodesmus* to embrace all such smooth-walled, monospinous desmids, whether bi-, tri- (or more) radiate (Fig. 13: 1, 3, 4). He emphasized in his study that cell shape in front view was a much more significant taxonomic character than shape as seen in the apical view (Teiling 1967, but see however, Bicudo 1975).

Similarly, in an attempt to solve some of the anomalies of the unwieldy and artificial genus *Staurastrum*, Palamar-Mordvintseva (1976) has proposed three new genera. The first is *Cylindriastrum* whose triangular recto-cylindrical cells are typified by the desmid originally called *S. pileolatum* (Fig. 11: 2). Second is *Cosmoastrum* with cells *Cosmarium*-shaped in outline whose walls are covered with short spines as in *S. polytrichum* (Fig. 13: 5). The third proposal is for a genus *Raphidiastrum*, with smooth walls as in the genus *Staurodesmus*, but differing in that the angles instead of bearing a single spine, bear two or three as in *S. brasiliense*, the type species (Fig. 11: 4).

Two other genera have been established in which the cells are polygonal in apical view. One is *Euastridium* which as its name implies basically resembles *Euastrum*. Its four species differ significantly in that the semicells when viewed both frontally and apically can be seen to bear rings of apical and lateral lobes. Also the numbers of these upper and lower lobes which range from 3 to 6 and 6 to 9 respectively always differ (Fig. 6: 5).

The monospecific genus *Amscottia*, described by Grönblad (1954), from Brazil is remarkable in that, like *Allorgeia* (see p. 13), adjoining semicells are distinctly different. It is a *Staurastrum*-like genus in that one of the globose semicells, in front view, has a projecting apex and two series of superimposed, slender processes which curve towards the apex. The other semicell has a rounded apex, but the two series of superimposed processes curve in the opposite direction and thus towards the apex of the adjoining semicell (Fig. 7: 2).

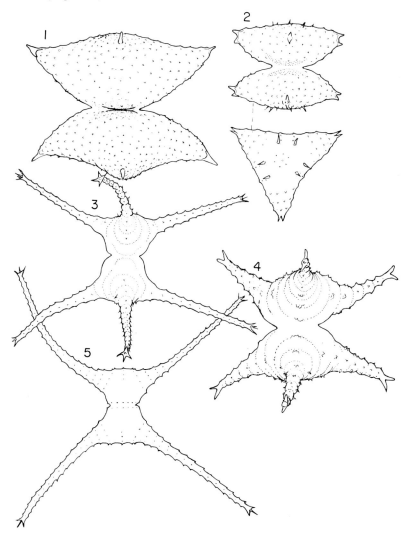

Fig. 12. Placoderm desmids of the suborder Desmidiinae: 1 *Staurastrum lunatum* var. *planctonicum* W. & G. S. West; 2 *S. avicula* Bréb.; 3 *S. pingue* Teil.; 4 *S. anatinum* var. *truncatum* West; 5 *S. chaetoceras* (Schrod.) G. M. Smith.

C

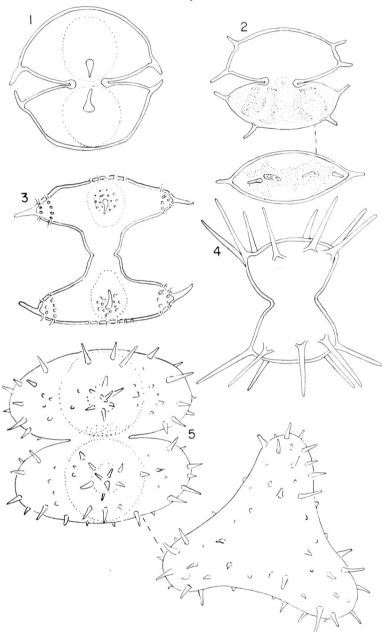

Fig. 13. Placoderm desmids of the suborder Desmidiinae: 1 *Staurodesmus dickei* var. *circulare* (Turn.) Croasdale; 2 *Arthrodesmus mucronulatus* Nordst. (after Bourrelly); 3 *Staurodesmus cuspidatus* var. *inflexum* (Racib.) Teiling; 4 *S. wandae* (Racib.) Bourr. (after Bourrelly); 5 *Staurastrum* (*Cosmoastrum*) *polytrichum* Perty.

The filamentous habit has also been adopted by some desmids which are polygonal in apical view. Thus the cells of *Desmidium*, which in apical view, depending on the species, can be seen to be triangular, quadrangular or even pentagonal, adhere together in long filaments which often show a helical twisting and are embedded in a gelatinous sheath. In some species the cells are joined by their apices but in others, the cells possess polar mamillae so that there is a space separating the cells when joined by such appendages (Fig. 14: 2).

Although the cell morphology of the filamentous *Phymatodocis** may appear insufficiently different to warrant such forms being placed in a genus separate from *Desmidium*, their mode of cell division is so distinctive that such generic separation is fully justified (see Chapter 5). However, like some species of *Desmidium*, the cells are quadrangular in side view with a deep median construction. In vertical view they appear irregularly quadrate but with a broadly rounded arm at each of the four corners.

Another filamentous genus with an unusual mode of cell division is *Bambusina* (= *Gymnozyga*) which includes about a dozen species. Some have cylindrical cells, others are reminiscent of *Staurodesmus* in that their octagonal cells bear single spines at the angles. The cells are united to form long straight filaments usually enclosed in a gelatinous sheath. Cell division as in *Phymatodocis* is unusual for desmids, resembling that seen in some species of *Spirogyra*, in that they show the formation of an annular infolding prior to cell elongation (see Chapter 5) (Fig. 14: 1).

The filamentous genus *Streptonema* with only one species, *S. trilobatum*, found in India and Borneo (Scott and Prescott 1961) also shows a similar annular infolding prior to cell division. The filaments are composed of large cells, each with a deep median constriction and rounded angles. In apical view the semicells appear like trefoils with three swollen lobes. The cells are joined by three slender apical appendages which arise from the narrow region of each arm of the trefoil and are held together by gelatinous buttons (Fig. 14: 3).

The cells of the filamentous *Groenbladia* (Teiling 1952) are cylindrical and similar to those of *Hyalotheca*. They are distinguished, however, by the possession of a very characteristic chloroplast which is an axile ribbon containing from 2 to 4 pyrenoids, and reminiscent of *Mougeotia* or *Roya* (Fig. 10: 2).

The presumed phylogenetic relationships of desmids, with their incredible range of morphologies, have been discussed by various authors (Fritsch 1935, Palamar-Mordvintseva 1976, Mollenhauer 1973) but the actual relationships are purely conjectural. Aspects of the possible causes

*Lorch (pers. comm.) has shown the cell wall to be so distinctive that this genus must now be placed in a separate family.

and consequences of their evolution are, however, enlarged upon and discussed in Chapter 2, which deals with the concepts of radiation, symmetry and asymmetry, and in Chapter 3 on the internal structure of the desmid cell and especially the chloroplast.

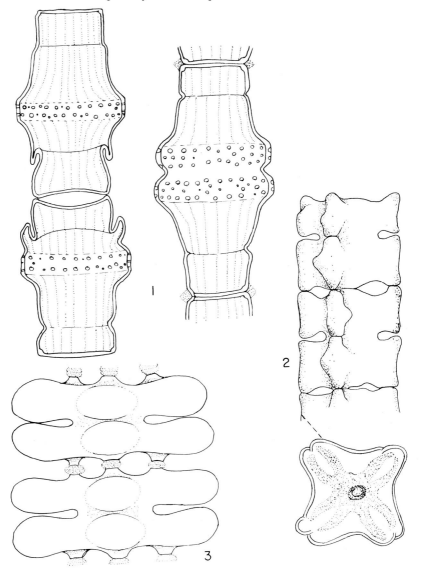

Fig. 14. Placoderm desmids of the suborder Desmidiinae: 1 *Bambusina brebissonii* Kütz. (after Bourrelly); 2 *Desmidium baileyi* (Ralfs) Nordst.; 3 *Streptonema trilobatum* Wallich (after A. M. Scott).

CHAPTER 2
RADIATION, SYMMETRY AND
ASYMMETRY

From the foregoing survey of the range of desmid morphology it should be apparent that one of the unique features of these algae, at least in the case of the placoderm desmids, is their often elaborate, polysymmetric structure. Each placoderm cell is divided, either by a suture line or a more or less prominent isthmus, so that adjacent semicells are symmetrical and almost perfect mirror images of one another (Fig. 15: 1A, 2A, *a-a'*). When still observed in front view the shape of each cell lying on either side of a line passing from semicell apex to apex through the centre of the isthmus is symmetrical, so that there is a second plane of symmetry about the central axis (Fig. 15: 1A, 2A, *b-b'*). Also there is a plane of symmetry perpendicular to the vertical axis so that except when the cells are cylindrical in apical view, they appear when seen from this position elliptical, triangular, or as having higher orders of angularity (Fig. 15: 1B, 2B, 3, 4). The corners of the angles may be rounded, but, especially in desmids with a basically triangular polygonal shape, the angles may be considerably extended to form ray-like processes, as previously described with respect to *Staurastrum* (p. 20 and Figs 12: 1–5, and 13: 3, 4). Hence desmids with this type of morphology are said to be *radiate* and to possess particular degrees of radiation, and a cell with three corners is termed triradiate (Fig. 15: 2A, B), one with four, quadriradiate (Fig. 15: 3) and with five, pentaradiate (Figs 11: 4 and 15: 4). Radiations of 5–12 are summarized as pluriradiate (Figs 6: 5 and 16: 1). Elliptical cells with only two corners which may be seen in most species of *Cosmarium* (Fig. 5: 4), *Euastrum* (Fig. 6: 1–3) or *Micrasterias* (Figs 6: 7 and 7: 1), or which as in the case of some species of *Staurastrum* may in addition bear processes, are termed biradiate.

   Clearly biradiate desmids have two perpendicular, vertical planes of symmetry which when seen in apical view form the axes of a more or less flattened ellipse (Fig. 15: 1B, *c-c'* and *d-d'*). Most species of *Staurastrum* have three or more planes of symmetry at 120° (Fig. 15: 2B) or less, to one another, this angle depending on the number of radiate processes they possess (Fig. 15: 3, 4). It should, however, be pointed out in this connection

27

that the extent of a desmid's radiation must not be uncritically identified with the number of processes borne on a semicell. For example, the fairly common planktonic *Staurastrum arctiscon* has 15 processes (Figs 16: 2 and 17: 1), but it is a triradiate and not a 15-radiate desmid. Similarly

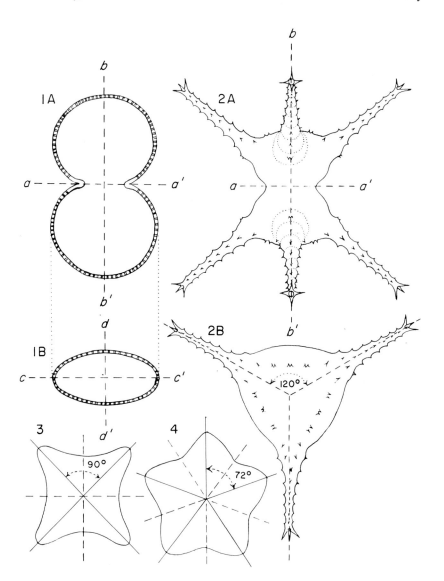

Fig. 15. 1–4 Planes of symmetry in placoderm desmids. (For explanation see text, p. 27.)

1

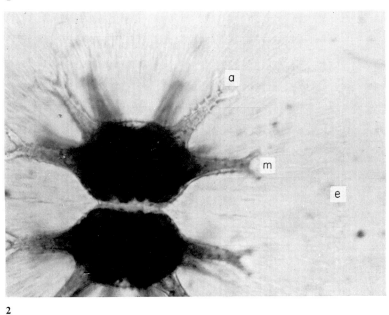

2

Fig. 16. Radiation in placoderm desmids: 1 apical view of a semicell of the polyradiate cell of *Staurastrum ophiura* Lund. This species varies in the degree of its radiation from 4- to 9-radiate; 2 Lateral view of *S. arctiscon* (Ehr.) Lund., showing lower major processes (m) and accessory ones (a) (cf. Fig. 17: 1, 2). Note the enveloping mucilage sheath (e).

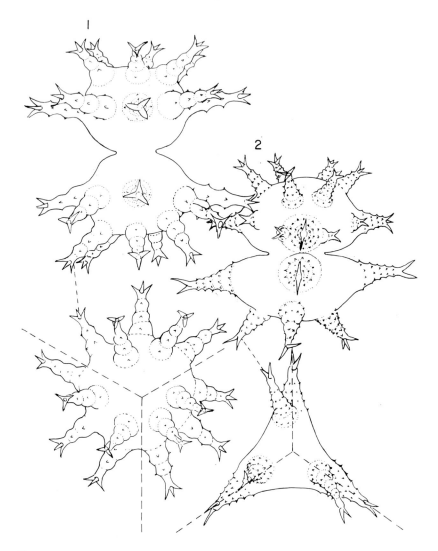

Fig. 17. Symmetry in placoderm desmids: 1 *Staurastrum arctiscon* (Ehr.) Lund., in side 'a', and vertical 'b' view. The latter shows that although there are 15 apical processes, this desmid is in fact, triradiate; 2 *S. furcigerum* Bréb.—a dichotypic specimen which, despite the numerous accessory processes in the upper semicell, is also triradiate.

S. furcigerum (Fig. 17: 2) is also triradiate, but each of the three major processes bear one or two accessory ones, so that specimens appear to be either 6- or 9-radiate.

Cells which are circular in end view, as are all saccoderm desmids as well as such placoderms as *Penium, Closterium, Pleurotaenium* etc., and some species of *Cosmarium*, are said to be omniradiate with an infinite number of vertical planes of symmetry.

In the table below, modified from Teiling (1950), the range of radiation encountered in the desmids is summarized (Table 1).

Only three algologists, Teiling (1950), Fritsch (1953) and Mollenhauer (1975b), have dealt in any detail with radiation and its implications for desmid taxonomy and evolution. Prior to their papers, however, various

Table 1. A summary of the range of radiation encountered in various genera (after Teiling 1950).

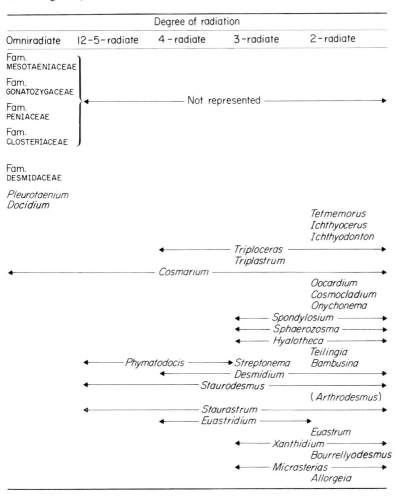

authors had examined aspects of the phenomenon with respect to various taxa in natural population (Heimans 1942, Turner 1922, Playfair 1910) and in cultures (Lefèvre 1939).

Teiling (1950) summarizes this fascinating aspect of desmid morphology as follows:

1 Radiation is unique to desmids and no other algae (except a few with spherical cells) are symmetrical in three perpendicular planes.

2 Radiation is a variable character, but the extent of such variation differs. The biradiate condition is the most constant form and there seems to be increasing variability with increasing degrees of radiation.

3 Some species appear to occur exclusively and constantly in certain biotopes in the triradiate form; in others in the biradiate condition; in others there may be an admixture of the two expressions of radiation.

4 A few species have been observed which show a recurring seasonal shift in their radiation from bi- to triradiate, and then back to biradiate (Reynolds 1940) (see pp. 33 and 238).

In considering the evolutionary implications of the range in the degree of radiation found in desmids the earlier students of these algae (West and West 1904, Fritsch 1933, Prescott 1948) accept the notion of a phyletic series from omniradiate→biradiate→triradiate→pluriradiate. Teiling (1950) argues, however, that evolution proceeded in a contrary direction, and as has occurred in many examples in both plant and animal kingdoms, 'evolution has run its course (in desmids) from manifoldness to few'. Consequently he considers the biradiate genera to be the most highly evolved and argues that they have passed through what he terms 'reductive evolution' so that the genera *Cosmarium*, *Xanthidium*, *Euastrum* and *Micrasterias* are the most highly evolved. He also suggests that the triradiate condition, which some species of these genera produce, are reversions or atavisms. Table 1 is modified from a diagram in his 1950 paper and summarizes his views about the evolutionary development of radiation in different desmid genera.

Commenting on another diagram (his Fig. 34) which succinctly presents his well reasoned views about the radiation and phylogeny of the major placoderm desmid genera, he suggests that the 'richness in radiation of the staurastroid desmids may be due to their being a younger, or retarded, series of evolution'. Also he points out that primitive characters, under certain ecological conditions may have survival value in certain combinations. Thus the anguloradiate desmids, and especially those with extended processes, have an opportunity of presenting a greater surface to the light, especially when, as many desmids do, they grow in shaded aquatic environments among other algae, or in association with mosses or aquatic macrophytes which may greatly reduce the incident light intensity. Thus anguloradiate desmids, such as *Staurastrum* spp., are less dependent on the position in which they happen to be in relation to the direction of incident light than leaf-like, biradiate forms such as *Micrasterias*. The latter are clearly unfavourably placed to photosynthesize if turned edge-on to the incident light. This advantage is presumably compensated for, at least in part, by the ability of many of these desmids to move phototactically (see p. 67).

An additional ecological advantage of the radiate condition, apparent in many planktonic staurastra, is that many species common in lake plankton have very long processes, thus increasing their resistance to sinking (see p. 204).

Few experimental studies have been carried out on any of the increased surface to volume ratios associated with increased radiation and their presumed ecological–physiological significance, especially in relation to their photosynthetic efficiency.

ASYMMETRY

As in the case of all organisms which exhibit symmetry, so with desmids, exact symmetry in any plane rarely, if ever, exists. There are usually small differences in certain dimensions, ornamentation, development of processes, though they can often be recognised only by careful scrutiny. Such minor differences do not, however, alter the overall impression of symmetry with respect to all the planes mentioned above.

However, desmid cells are found, not uncommonly, in which striking differences are apparent in the shape and/or ornamentation of adjacent semicells so that they are markedly asymmetrical with respect to one or more planes. Many cases of asymmetry are recorded and illustrated in the literature (Figs 7: 1, 2 and 8: 1). Some are constant characters and possibly genetically determined though the majority are undoubtedly developmental (phenotypic) in origin and result from environmental factors being different when new semicells are forming from those prevailing when the older, parent cell was formed (see pp. 108–117). Because it is a phenomenon of great morphogenetic interest and significance, this aspect of asymmetry will be considered in some detail in Chapters 7 and 8.

This *vertical asymmetry* seems to be encountered most frequently amongst populations of planktonic desmids and here some species of radiate desmids exhibit seasonally, distinct differences in morphology between adjacent semicells of a given cell (Reynolds 1940). Thus in *Staurastrum chaetoceras* (Fig. 18) Reynolds (1940) incorrectly named it *S. paradoxum* (Brook 1959b) one semicell may be biradiate, the other triradiate. In some species the differences in radiation may be even more striking.

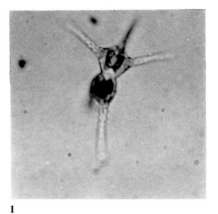

1

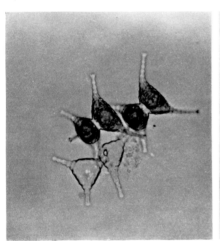

2

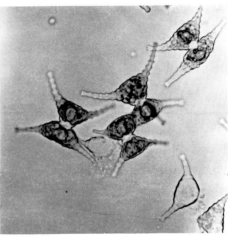

Fig. 18. Asymmetry in placoderm desmids: 1 a dichotypic specimen (Janus 2 + 3) of *Staurastrum chaetoceras*. The upper semicell is triradiate, the lower biradiate; 2 Specimens of the same species from laboratory-maintained cultures. Note the shorter processes of these specimens.

If the differences are not teratological, as they often are especially in laboratory cultures, these so called *dichotypical* specimens (Teiling 1948) may suggest relationships between the desmid in question and varieties of it, or even with different species, and in extreme cases genera. Because of their 'two-headed' form Teiling has named such specimens as 'Janus' forms, after the ancient Roman god presiding over doors and gates and usually represented with two heads looking in opposite directions. Janus forms are most frequently encountered in species of *Staurastrum* and *Staurodesmus* (Fig. 19: 1, 2) though they also are to be seen in species of

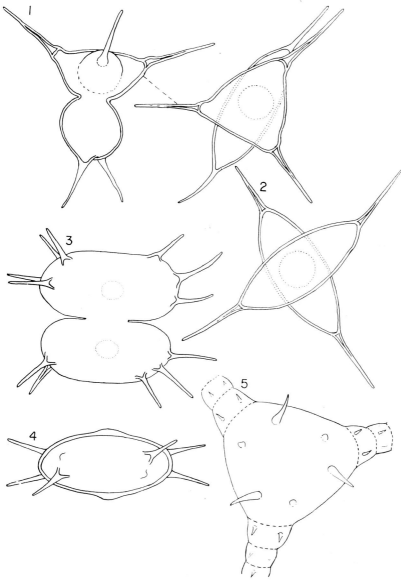

Fig. 19. Asymmetry in placoderm desmids: 1 lateral (left) and apical (right) views of a dichotypic specimen (Janus 2 + 3) of *Staurodesmus sellatus* Teiling; 2 apical view of Janus 2 + 2 (biradiate) specimen of the same species showing torsion asymmetry of adjacent semicells; 3 and 4 lateral asymmetry in *Xanthidium subhastiferum* West. 3 is a side view, and 4, an apical view which shows clearly the unequal production of the apical spines; 5 apical view of a semicell of *Staurastrum pingue* Teiling showing asymmetric development of apical spines; an example of semiradial asymmetry.

Xanthidium, Micrasterias and *Cosmarium* (*Cosmostaurastrum*). The examination of plankton collections from many hundreds of lakes in Great Britain and the mid-west of North America (Brook 1959a, b, 1966, 1970) indicates that the following planktonic species exhibit dichotypic forms: *Staurastrum anatinum, S. cingulum, S. gracile, S. chaetoceras, S. manfeldtii, S. tetracerum, S. pingue, S. planctonicum, Staurodesmus cuspidatus, S. curvatus, S. dejectus, S. jaculiferus, S. joshuae, S. megacanthus, S. mammillosus, S. sellatus, S. subtriangularis, S. triangularis* and *Xanthidium subhastiferum.*

Teiling (1950) has suggested a notation for recording Janus forms such that (2 + 3) would indicate a biradiate and adjoining triradiate cell; (3 + 4) a triradiate and quadriradiate cell. An analysis of collections made by Teiling of 33 examples of Janus forms in 25 species of desmids produced the results in Table 2.

Table 2. Analysis of occurrence of Janus forms (after Teiling 1950).

Janus form	(2+3)	(3+4)	(4+5)	(2+4)	(3+5)	(3+6)
No. of cases	15	9	3	2	3	1
	Janus difference = 1			Janus difference = 2		Janus difference = 3
Total examples	27			5		1

From this it is clear that the most frequent degree of difference in radiation between semicells is one, though even out of such a limited sample a significant proportion of greater differences do occur.

Many investigators of desmid morphology when describing and, unfortunately, putting names to single specimens rather than examining populations, have given specific or at least varietal rank to forms which differ only in their degree of radiation. Thus *Staurastrum thummarkii* undoubtedly is merely a biradiate expression of *S. cingulum* (Brook 1959b). Similarly many so called species of *Arthrodesmus* are biradiate forms of what were considered to be monospinous species of *Staurastrum* until Teiling (1948) erected the genus *Staurodesmus. Xanthidium antilopeum* is normally biradiate but from some lakes (e.g. Grasmere and Easdale Tarn in the English Lakes) triradiate forms have been described and named as var. *triquetia* Lundell; similarly *X. subhastiferum* West is most commonly biradiate, but there is a var. *murrayi* which though mostly biradiate does produce triradiate cells especially in the plankton of Loch Lomond. Such cells have, however, only been afforded the taxonomic rank of forma; that is forma *triquetra* West and West (Fig. 20).

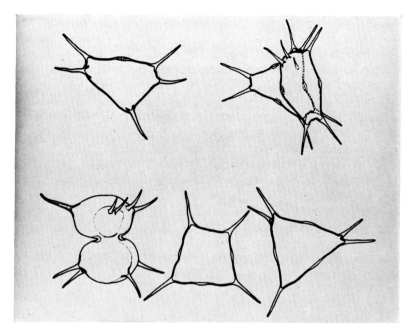

Fig. 20. Asymmetry in placoderm desmids. Cells of *Xanthidium subhastiferum* var. *murrayi* West from the plankton of Loch Lomond, Scotland. Biradiate, triradiate and even quadriradiate cells may be encountered, including dichotypic specimens as shown in this figure.

As pointed out by Teiling (1950) there are no terms or rules of nomenclature to cope with desmids with variable radiation, and as indicated above, sometimes they are distinguished as varieties, sometimes only as forms. Indeed the concepts of *varietas* and *forma* are used in desmid taxonomy to signify deviations, of greater or lesser degree, from the description of type species which it is almost certain in the case of some has been made from only one, possibly quite atypical specimen, rather than being a description of the range of characters encountered in a population. Thus the use of variety and forma has been quite capricious, and a general and very unscientific tradition would seem to be that the first deviation from the type species to be described is regarded as a variety, no matter what the character or extent of the variation is. Any, deviations found later are indicated by *forma*.

Fortunately, there has been a more rigorous trend in desmid taxonomy so that the range of morphology exhibited by a population is studied and recorded. Such studies have led to the realization that many of the varieties and forms described as such in the literature have no taxonomic validity and must therefore be rejected. Valuable studies in this connection which have begun to clear up some taxonomic inconsistencies are, for example, those of Brook (1959b, c), Brook and Hine (1966), Lind and Croasdale (1966), Růžička (1973) and Tassigny (1966).

Because it may be important to record the extent of radiation in desmid populations especially in relation to their ecology, Teiling (1950) proposed the term *facies* (abbreviated to *fac.*) for use as a nomenclatural expression to record this morphological feature. Being independent of variety or form, the facies terminology can be placed after the designation *varietas* or *forma*. Thus in the cases instanced above, *X. antilopeum* var. *triquetra* becomes *X. antilopeum* fac. *triquetrum* and *X. subhastiferum* var. *murrayi* forma *triquetra* becomes *X. subhastiferum* var. *murrayi* fac. *triquetra*. It was for similar sound reasons that Teiling (1950 and see p. 34 above) proposed the adoption of 'Janus' as a nomenclatural expression to be used exclusively to indicate the character of specimens which are dichotypic with respect to radiation. Thus specimens of *Staurastrum chaetoceras* Janus (2 + 3) would be those individuals which had adjoining biradiate and triradiate semicells while *S. chaetoceras* fac. *biradiata* and fac. *triradiata* would indicate completely biradiate or triradiate forms respectively. These proposals have been adopted by a few algologists.

TYPES OF ASYMMETRY

Univertical Asymmetry

The most intriguing type of vertical asymmetry from a morphogenetic point of view has been discovered in the past 25 years in several tropical desmids. It is so distinctive and yet so constant in its expression that new genera have had to be established for them, e.g. *Amscottia*, *Allorgeia*, etc. (see p. 13 and Figs 7: 1, 2 and 8: 1). It will be recalled that in these desmids, the processes of one semicell are vertically curved, but those of the adjoining one curve in a contrary manner, so that all the processes are curved in one direction. In the *Micrasterias*-like *Allorgeia* the asymmetry is even more complex between adjacent semicells (p. 13). This expression of asymmetry has been termed *univertical* asymmetry (Teiling 1957).

Torsion asymmetry

When the processes of one semicell do not lie directly above or below those of the other, so that the cell appears to be twisted at the isthmus (a feature characteristic of several species of *Staurastrum*), such asymmetry is termed *torsion asymmetry*. Indeed *S. alternans* is so called because this is a constant diagnostic feature of the species. Torsion asymmetry is also characteristic of the almost ubiquitous plankton desmid *S. pingue*, and also some species of *Staurodesmus* from the plankton (Fig. 19: 2).

Lateral asymmetry

This type of asymmetry is not uncommon in species of *Xanthidium* with respect to its apical spines, and especially in *Micrasterias*, where it may be seen in some species in their elaborately indented apical processes. Their asymmetry is caused either by a doubling, or a lack of production of spines, or processes, and has been studied in detail in *Xanthidium subhastiferum* (Fig. 16: 2) by Rosenberg (1944) and *M. mahabuleshwarensis* f. *wallichii* (Fig. 6: 7) by Teiling (1956).

Corporeal asymmetry

Corporeal asymmetry is restricted to aspects of the body of the semicell rather than to radiation or ornamentation. The most striking examples are to be found in the genus *Closterium* (p. 8) a genus in which the vast majority of species possess only one vertical symmetry plane. Several species exhibit an asymmetrical, spirally turned or sigmoid shape within populations of the normal, regularly curved specimens. In *Closterium acutum* var. *variabile* however, irregularities of this type seem to be a characteirstic feature of all cells of this desmid (Fig. 21: 5).

The colonial *Oocardium stratum* (see p. 16) is asymmetrical in a lateral plane and, as suggested by Teiling, the consequent exposure of a greater lateral area to light may be of ecological significance to this stream-living desmid in its calcareous tubes (Fig. 9: 1, 2).

In his discussion of corporeal asymmetry Teiling mentions the case of the unique *Cosmarium*-like desmid currently called *C. obliquum* Nordstedt. This authority (Nordstedt 1873) noted that its cells were consistently asymmetrical in all the material examined from Norway and Sweden, an observation confirmed by subsequent collections from diverse regions of the Northern Hemisphere. So distinctive is its asymmetry that Raciborski (1889) established for it the genus *Nothocosmarium* but this has not been adopted, though Cedergren (1932) emphasises that its features merit generic distinction. Triradiate forms (fac. *trigonium*) of this unusual desmid have been found in several localities. Although Sampaio (1949) for this reason then assigned the desmid to *Staurastrum*, this is also inappropriate (Teiling 1957) because the cell's unique asymmetry appears to be very constant. The latter author also points out that *Staurastrum cordatum*, which is asymmetrical in one lateral plane, possesses a structure which suggests that it too should be included in Nordstedt's genus *Nothocosmarium*.

Semiradial asymmetry

This type of asymmetry is potentially capable of occurring in almost any

D

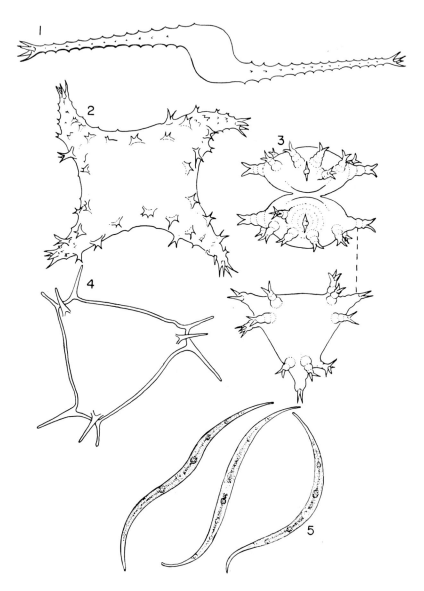

Fig. 21. Asymmetry in placoderm desmids: 1 semiradial asymmetry in *Staurastrum leptocladum* var. *sinuatum* G. M. Smith (after Smith); 2 apical view of *S. cyrtocerum* Bréb. showing inflexed radial processes; 3 Lateral and apical views of *S. furcigerum* Bréb. showing the lateral asymmetric development of accessory processes; 4 apical view of *S. trifidum* var. *inflexum* showing the asymmetrical development of one spine on each lateral face of the semicell; 5 corporeal asymmetry in *Closterium acutum* var. *variabile* (Lemm.) Kreig. (after Hortobagyi).

of the anguloradiate desmids no matter what their degree of radiation, from biradiate upwards. What is implied by this form of asymmetry, which can only be recognised when cells are seen in apical view, is that one side of each process has either an accessory process, spine, verruca or other type of ornamentation which is lacking on the opposite side. When turning the radii round the vertical axis it can be easily established that the radial processes are congruent but that each is asymmetric with respect to one or more morphological feature. Examples are illustrated in Fig. 21. Many genera exhibit this type of asymmetry which may be expressed with respect to a number of characters. Thus the filamentous genera *Onycho-nema* and *Sphaerozosma* show it in the disposition of their small apical processes (Figs 10: 4), primarily four, but which are commonly reduced asymmetrically to two. Species of *Euastrum* may show an asymmetric development of the apical lobes; and populations of *Staurastrum pingue* may have only the right verruca of each of the characteristically three apical pairs developed into a single spine. Nygaard (1949) gave this form varietal rank (var. *tridentatum*) but it seems doubtful whether this is warranted since it has been found not infrequently amongst 'normal' forms in plankton populations in British lakes (Brook 1959b) (Fig. 19: 5).

Semiradial asymmetry may also occur in the front face of some desmid cells and thus can be seen in both apical and front view and also seems to be a specific character. Thus *Cosmarium onychonema* has one tubercle situated to the right of the front of its semicells. *Euastrum pingue* has on its front face to the left, a granulated protuberance, and to the right, a scrobiculation. Similar examples have been described in the literature.

A very uncommon, but none the less interesting, type of semiradial asymmetry occurs when the radial planes are displaced in such a manner that they do not meet in the vertical axis. In some cases the displacement is barely perceptible but in others, as for example *Staurastrum leptocladum* var. *sinuatum* G. M. Smith, it is very obvious (Fig. 21: 1).

Many *Staurastrum* spp. have been described with curved or inflexed radial processes as a character of taxonomic significance. It is of interest to note that Teiling (1957), when examining this type of asymmetry, found that although the direction of the curvature may be to the right or left, 18 out of 24 examples curved consistently to the left. Very few species (e.g. *Staurastrum cyrtocerum* (Fig. 21: 2)) curve either to the right or left. The significance of these interesting observations has still to be explained.

A final type of semiradial asymmetry occurs when one side of a process in a radiate desmid is more developed than the other, and carries extra processes, or spines, which the other side lacks. It is a not uncommon, constant characteristic of a considerable number of *Staurastrum* species. For example, the right side of the radial angle of *S. trifidum* var. *inflexum*

(Fig. 21: 4) is more developed, with a longer spine displaced towards the middle of the front side; also the sub-apical spine is slightly turned to the right. This type of asymmetry is also well exemplified by the complex semicells of some varieties and forms of *S. furcigerum* (Fig. 21: 3) and more particularly by *S. sexangulare*. For other examples see Teiling (1957).

To summarize the main characteristics of semiradial asymmetry, Teiling (1957) offers the following generalizations about its occurrence and significance:

1   It usually occurs in both semicells, but may be confined to only one (i.e. it may be dichotypic).

2   It may be manifested by one or more details of external morphology; so far cases of internal asymmetry are unknown.

3   It affects one, or several characters, independently of others.

4   It is usually expressed as an enlargement, or reduction, or total absence of radii or emergences; or even parts of the latter.

5   It can play an important distinguishing role in taxonomic studies.

Although some aspects of the morphogenesis of asymmetry have been investigated in the excellent researches, especially of Waris and Kallio (1964), with particular reference to species of *Micrasterias* (see Chapter 8), much waits to be discovered about the factors which may underlie its expression not only in this genus, but in many other desmid genera. Such research could well prove to be of general applicability to the whole problem of cell development and differentiation.

CHAPTER 3
INTERNAL ORGANISATION OF THE
DESMID CELL

CHLOROPLASTS: STRUCTURE AND SIGNIFICANCE

As autotrophic organisms, desmids depend for their nutritional survival on the capture of light energy by the chlorophyll-containing organelles within their cells, the chloroplasts. Like other green algae these contain in addition to the chlorophylls 'a' and 'b', carotenes and xanthophylls.

A TLC examination of 33 distinct taxa of the class Zygnemaphyceae including both saccoderm and placoderm desmids has been undertaken by Weber (1969) who found that the Zygnemataceae (e.g. a *Mougeotia* sp.; 3 *Spirogyra* spp. and 2 *Zygnema* spp.) and Mesotaeniaceae (*Mesotaenium caldariorum* and *Netrium digitus* and *N. interruptum*) contain pigments identical to those of higher plants. The saccoderm *Roya obtusa*, however, and all placoderms investigated were found to contain an additional and unusual carotenoid. His analysis indicated the pigment to be a tri-hydroxy--carotene, and he suggests that this unique pigment may be of significance in the biochemical delimitation of the Desmidiaceae from other green algae. Weber also presents data on chlorophyll 'a', 'b' and carotenoid ratios (see also Drawert 1966, Hager and Meyer-Bertenrath 1966 and Herrman 1968).

The initial studies of desmid chloroplasts by Lütkemüller (1893, 1895) and especially Carter (1919a, b, 1920a, b) indicated that in many cases they are very elaborate structures requiring careful and critical microscopic examination to understand and elucidate their true form. More recently attention has been directed to their submicroscopic structure by various investigators (Drawert and Mix 1961, Chardard 1972, Wartenberg and Dorscheid 1964) and these studies have indicated considerable complexity in their fine structure.

CHLOROPLAST TYPES

Seen at the level of magnification of the light microscope two major types of chloroplasts can be recognised: *axile* ('situated in the centre or axis') and *parietal* ('pertaining to the wall'). Thus an axile chloroplast, in the strict sense, refers to one with most of its chlorophyll round the vertical

axis where also the pyrenoid(s) are normally situated. A parietal chloroplast has most of its photosynthetic mass and pyrenoids in the distal part
of the cell adjacent to the cell wall. These categories are by no means clear
cut and it must be stressed that elaborate intermediate forms exist in which
basically axile chloroplasts have much of their photosynthetic surface in
a parietal position. Indeed, the truly parietal chloroplast is rare and consists of several plates or ribbons (very exceptionally a longitudinal cylinder)
which, with their pyrenoids, are situated round the periphery of the cell.

A suggested phylogenetic classification coupled with a reasoned
explanation for the vast range of shapes exhibited by desmid chloroplasts,
in association with the complexities of cell morphology, has been offered
by Teiling (1952). He postulates that the ecological factor light has been
the most significant in determining desmid evolution. Light, he suggests,
has been the principal cause of both cell extension and flattening to
produce the biradiate form and with this there has been a concomitant
elaboration of chloroplast structure. This has been coupled with the
ability to move phototactically thus enabling cells to take up photosynthetically favourable positions (see p. 67). Teiling thus proposes that
there has been an enlargement, in the course of evolution, of the chloroplast in both an apical and radial direction, hence placing the light-
capturing chlorophyll in a more advantageous position by its shift from
the axile to parietal position close to the cell wall, and through an increase
in its surface area by the development of ridges, papillae and marginal
extensions.

Teiling (1952) suggests that the axile *stelloid chloroplast* is the most
primitive (Figs 22A and 23) and indeed in all the saccoderms the chloroplasts are basically, though not necessarily obviously, of this type. The
typical stelloid chloroplast has an axile core with one or more embedded
pyrenoids and from the core there extend a number of longitudinally
radiating lamellae. These are found in the genera *Cylindrocystis* and
Netrium but also in what are assumed to be the simpler placoderms,
Penium and *Closterium*. Also in the higher placoderms, which have an
isthmus and so belong to the Cosmariae, the primitive stelloid condition
is found in the genera *Docidium* and *Pleurotaenium* and in a small assemblage of more or less elongate cylindrical *Cosmarium* spp. of the *clevei*
group. In the saccoderms *Mesotaenium*, *Ancylonema* and *Roya* the
chloroplast is described as an axile ribbon though as emphasised by
Wartenberg and Dorscheid (1964) in *Mesotaenium violascens*, the underlying structure is said to be helicoidal. Ribbon-like chloroplasts in these
forms, according to Teiling, are derived by reduction from the stelloid
type, and he categorises them as *laminate* chloroplasts (Fig. 22 I C, D & E).
Evidence for their origin from the stelloid type derives from the observa-

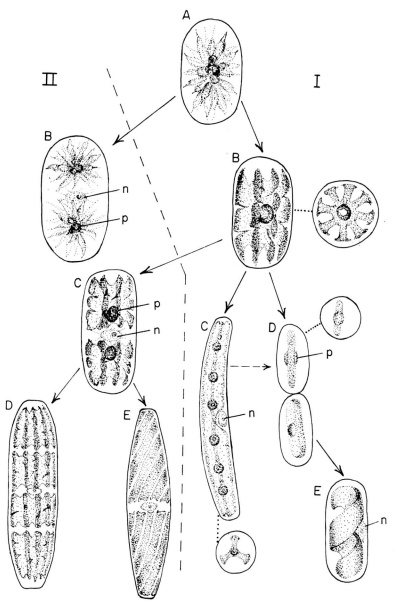

Fig. 22. The two postulated lines in the evolution of the desmid chloroplast from its presumed single stelloid ancestral type seen in *Cylindrocystis acanthospora* Krieg. (after Mollenhauer) (A). Line I is represented by the primitive saccoderms with undivided laminate chloroplasts viz. *Roya obtusa* var. *anglica* West (Krieg.) (C), *Ancylonema nordenskioldii* Bergg. (D) and the simple spirally twisted chloroplast of *Spirotaenia endospora* Bréb. (E). Line II indicates the presumed evolution of the more advanced saccoderms all of which exhibit 'precocious chloroplast division' (see text), now a stable character var. *major* West (B) *C. brebissonii* Menegh. (C). *Netrium interruptum* (Bréb.) Lütkem. (D) *Spirotaenia trabecula* A. Br. (E). n = nucleus; p = pyrenoid.

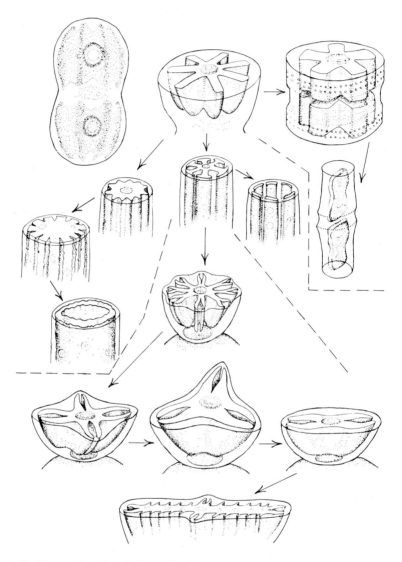

Fig. 23. Diagram based on Teiling's (1952) postulated evolution of the desmid cell from the omniradiate cell with stelloid chloroplast to, in the most advanced desmids (Cosmariae), the laminate-furcoid condition.

tion that under certain growth conditions cells of *Mesotaenium, Roya* and the placoderm *Gonatozygon* (see p. 8) will produce chloroplasts with a stelloid structure. According to Grönblad (1935) stelloid chloroplasts with few lamellae can sometimes be found in *Roya* and a similar condition condition has been seen in *Gonatozygon*.

An analogous reduction to the laminate condition would seem to have occurred in the filamentous placoderm genus *Hyalotheca*, also typified by cylindrical cells. Thus, although *H. dissiliens* has stelloid chloroplasts in each semicell, each of those of *H. neglecta* is said to be axile and laminate, so that West and Carter (1923) described it as 'axile', one in each semicell (Fig. 23), typically with a single central pyrenoid in each, and plate-like, so that when viewed from the edge it seems narrow just as in '*Mougeotia*'. Because of this distinctive feature, Teiling (1952) erected the new genus *Groenbladia* for it. Of very special significance with regard to Teiling's hypothesis concerning the stelloid origins of the laminate chloroplast is Grönblad's discovery (1921) of 'triradiate' chloroplasts in some forms of *H. neglecta*. As pointed out by Teiling, this discovery supports the view that laminate chloroplasts are derived from the stelloid by reduction.

In the most primitive saccoderms, although many do not now have truly stelloid chloroplasts, their reduced laminate ones usually extend from pole to pole of the cell so that the nucleus occupies a somewhat lateral position half way down the length of the cell (Fig. 22 I, C). As already indicated (p. 6) the chloroplast may be interrupted in the median region so that it has a distinct median notch (e.g. *Roya*). In the presumably more advanced cells (*Cylindrocystis* and *Netrium*) two chloroplasts are present in each cell, separated by a centrally placed nucleus. Mollenhauer (1973) has suggested that this tendency to divide the chloroplast prior to cell division (prazedente Plastidomatomie=precocious chloroplast division) has become a stable character in the course of evolution and now constitutes an essential feature of all but the most primitive desmids (Fig. 22 Line II).

The diagram in Fig. 23 summarizes Teiling's views on the evolution of the stelloid chloroplast from the postulated ancestral type still found, according to him, in *Penium* and *Cosmarium* spp. of the *clevei* group, to the parietal chloroplast, which is the highest expression of the evolution of this cell organelle.

As indicated above, one trend has been in the reduction of the numerous lamellae to two, so that a simple ribbon-like chloroplast results. Another proposed trend has been the incision of the lamellae at their apices, an important tendency which led the way to the truly *furcoid chloroplast* typical of the majority of the placoderm desmids. In this, there has been a reduction of the central axis either with a corresponding thickening of the edges of the lamellae, or they have become transversely flattened to such a degree that in their parietal position they appear on casual observation to be longitudinal ribbons. Indeed it was for this reason that all species of *Spirotaenia* were believed to possess ribbon-like twisted parietal chloroplasts, rather like those of *Spirogyra*, until Lütkemüller (1895) showed them in fact to be axile and stelloid with axile pyrenoids (Fig. 22 II, E). There are, however, some species of *Spirotaenia* of the sub-genus *Monotaeniae* that do have truly parietal chloroplasts and associated pyrenoids without a central axis or core (Fig. 22 I, E).

A reduction in the central axis can be seen in some species of *Closterium*, especially in the narrow apical regions of their cells, though the distal ends of the lamellae may be bi- or even trifurcate. In some other *Closterium* species there would seem to have been an evolutionary tendency towards an increase in the thickness of the central axis of the chloroplast, so that when seen in section, it appears like a cog wheel, though there is a parietal arrangement of the pyrenoids (Fig. 23).

However, the most important manner in which desmids would seem to have increased the area of chloroplast surface is by its radial enlargement, and this is exemplified only by placoderms of the tribus Cosmariae. This radial expansion of the lamellae is intimately associated with corresponding angular extensions of the cell wall initially as protrusions, or in their ultimate development, as extended processes. Increase in surface area of the lamellae has resulted from the longitudinal bipartition of their distal edges so that the lamellae appear fork-like (furcoid) when viewed from the cell apex (Figs 23 and 24). The bipartition may form such a deep cleft that the lamellae appear to be double structures. Teiling has termed these *semi-lamellae* to distinguish them from the laterally protruding secondary ridges or *cristae* which in very many biradiate desmids may constitute an important component of the chloroplast.

Teiling (1952) has outlined how the furcoid chloroplast may have gradually evolved and in so doing distinguishes a series of distinctive types. Thus the primitive type is exemplified by such pluriradiate species of *Staurastrum* as *S. ophiura* and *S. brasiliense* var. *lundelli*, both of which have long, well-developed processes. Short-armed, triradiate and thin biradiate desmids of moderate size (20–30 µm) possess lamellae which are completely split into two halves and the central core with an axile pyrenoid forms a robust centre to the chloroplast (Fig. 25). In convex-sided triradiate or thick biradiate cells, short secondary lamellae (cristae) develop from the core and extend outwards towards the cell wall. Larger forms of biradiate, and in a very few triradiate desmids, such as *Staurastrum anatinum* and *S. sexangulare*, exhibit a bi- or even tripartition of the core

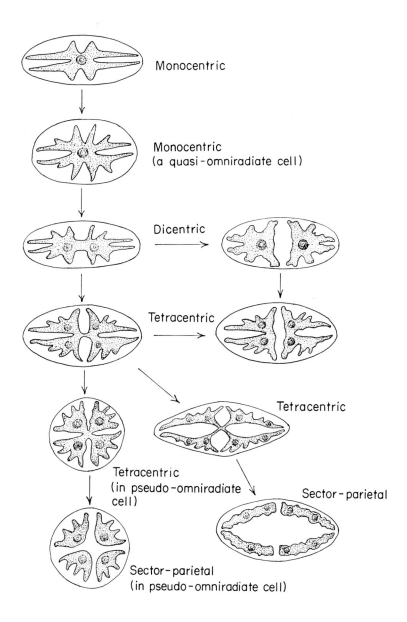

Monocentric

Monocentric
(a quasi-omniradiate cell)

Dicentric

Tetracentric

Tetracentric

Tetracentric
(in pseudo-omniradiate
cell)

Sector-parietal

Sector-parietal
(in pseudo-omniradiate cell)

Fig. 24. The probable evolutionary sequence from the monocentric, furcoid chloroplast through dicentric, tetracentric to the sector-parietal condition found in some large cells of members of the Cosmariae (after Teiling 1952).

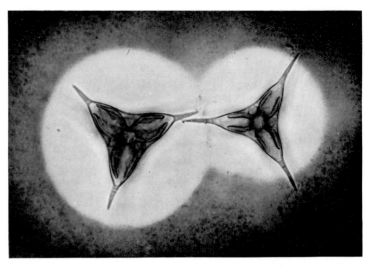

1

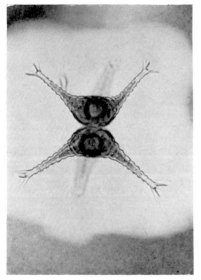

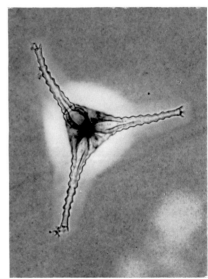

2

Fig. 25. 1 Apical view of the furcoid chloroplasts of two cells of the triradiate desmid *Staurodesmus curvatus* var. *inflatus* (Lind and Pearsall) Brook; 2 lateral and apical views of *Staurastrum anatinum* forma *paradoxum* Brook showing how the chloroplast extends into the slender, elongate processes. (Note: central pyrenoid in all cells and the mucilage envelope surrounding each cell.) (Photomicrographs by Dr Hilda Canter Lund.)

with pyrenoids and the central axis reduced to a narrow connecting strip lacking a pyrenoid. Many species of *Cosmarium* possess this *dicentric* type of chloroplast. Larger species of *Cosmarium* and the large cells of the genus *Xanthidium* show a further advance on this elaboration in that they possess four pyrenoids and there is a further reduction in the chloroplast core; or it may even be absent (Figs 24 and 26).

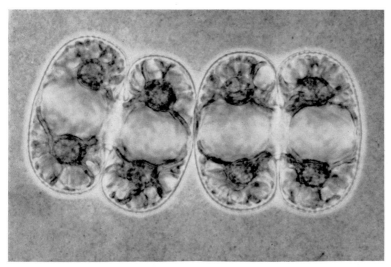

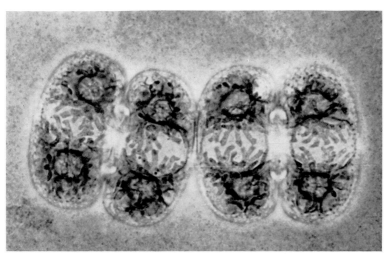

Fig. 26. Cells of *Cosmarium dentiferum* Corda at different levels of focus. In these cells the centre of the chloroplast is greatly reduced and the margins can be seen to be richly branched, so that their edges become ramified parietally, to form a complex fringe. (Photomicrographs by Dr Hilda Canter-Lund.)

What is assumed to be the final stage in this evolutionary progression can be seen in those cells in which the parts of the dicentric chloroplast are each divided further into two or more parts and their connecting core may remain in a considerably reduced form, or be non-existent. Each of the parts contains one or more pyrenoids.

Especially in large, elongate species of *Euastrum* there is a cross-shaped central connection containing a reduced primary pyrenoid in the middle. Such chloroplasts are functionally parietal even though a central core remains. They are found in biradiate desmids where the ratio of thickness to breadth is 1:2 or branched in front view and their edges become ramified parietally forming a dense confusion of irregular edges and fringes (Fig. 26).

The greatest degree of lateral compression of the desmid cell is in the genus *Micrasterias* and with it the furcoid chloroplast has been more or less lost. Instead it has become extended in two directions radially and the forked edges of the two lamellae are very small and rudimentary. The laterally extended lamellae bear ridges which in the thicker species may be quite long and extend to the cell wall: in the very flattened ones they are but squat strips (Fig. 23 bottom). Very large numbers of pyrenoids (up to 100) are to be seen in these cells. These very flattened cells of *Micrasterias* spp. have been thought of as analogous to the plate-like leaves of higher plants, and, because of the position of the chloroplast adjacent to the walls of these greatly compressed cells, they may be considered to be parietal in position. Their origins are, however, quite different from those of other parietal chloroplasts.

A final type of furcoid chloroplast is distinguished by the possession of radiating processes containing semi-lamellae with cristae, the former having pyrenoids in their distal ends. Such are termed *pericentric* by Teiling (1952) and, according to him, are the consequence of a peculiar mode of evolution in the disposition of the chloroplast and associated pyrenoids.

CHLOROPLAST TYPE IN RELATION TO CELL SIZE

As long ago as 1889 the Finnish botanist Elfving expressed the view that in small desmids the axile chloroplast was physiologically adequate because it could be easily penetrated by light. However with increasing size this became functionally less possible so there was an evolutionary response in a peripheral expansion of this organelle. Indeed it can be argued that as the axile chloroplast became large and more complex so that adequate amounts of light were prevented from reaching the core, the outer peripheral region developed at the expense of the inner. This centrifugal development has been expressed in different ways in cells with different morphologies. Moreover, that such a development from axile to parietal is at least in part a response to environmental conditions seems to be suggested by numerous observations. The large cells of *Staurodesmus grandis* provide a good example, for in the description in the Wests' British Desmidiaceae (West and West 1912) it is stated that the chloroplast is most frequently axile but may be parietal; moreover Carter (1920) found cells which were dichotypic in this respect, being axile in one semicell and parietal in the other.

Observations on how this state of affairs may arise had been made in the course of Carter's very detailed study of desmid chloroplasts (Carter 1919a,b, 1920a,b). In the last of these papers she followed the process of chloroplast regeneration in newly forming semicells after cell division. She found that in the flattened Cosmariae with a restricted isthmus the chloroplast enters the newly formed semicell as a more or less bilobed protuberance, while in triangular or pentangular cells, having normally 3- or 5-lobed chloroplasts, the latter enters the developing semicell from the older one, either as a 3- or 5-lobed mass, according to the parental condition. Thus the future form of the chloroplast is retained even during this early budding stage of its development (Figs 47: 5 and 51: 5, 6). In the axile chloroplasts of many species however, it was found that there is a distinct tendency for the peripheral parts to enter the new semicell much more rapidly than the central 'as if there were (was) an attempt on the part of the organism to cover the cell wall with photosynthetic material as quickly as possible no matter what happened to the interior of the cell'!! Such is the case in many species of *Staurastrum*, according to her observations, in which the angles of the young semicells may be largely filled by chloroplast even though the central part has barely entered. If division of the chloroplast at the isthmus was to occur under these conditions then several distinct chloroplasts would be produced in the newly formed semicell rather than one. It is this, Carter suggests, that happens in *Staurodesmus grandis* in which there is a not infrequent occurrence of parietal chloroplasts instead of the more normal single axile one. Moreover, she suggests that those species which typically possess parietal chloroplasts were originally derived in this manner.

Teiling (1952) stresses the importance of the ratio between the area of the transverse section of the desmid semicell and the area of the isthmal hole in relation to the formulation of daughter chloroplasts and to chloroplast evolution. Clearly there is a major structural difference between desmids with a comparatively small and hence restricted hole and un-constricted, or only slightly constricted cells. In the former category of desmids (most of the Cosmariae) the new chloroplast must be rebuilt by the rather amorphous elements of the mother chloroplast intruding through the isthmuthal aperture, while this does not happen in the latter category. Hence the chloroplast of the unconstricted desmids is a direct continuation of the mother chloroplast, in contrast to those of the constricted forms, which must be rebuilt as part of the process of each cell division. Clearly this is a cause of potentially greater plasticity and variability, and therefore probably an important factor in chloroplast evolution.

Since desmids tend to form isolated populations in lakes, ponds or bogs in which reproduction is largely asexual so that they are the result of successive divisions over many generations (see p. 100) it is easy to understand how such forms of chloroplast structure, if they were photosynthetically more efficient and hence of positive survival value, became established. On the other hand it is clear why when one considers that in so many desmid species 'every chloroplast is derived from an original bud of green material squeezed through a narrow passage' (the isthmus) one would expect a large proportion of abnormalities—and it would seem unwise (as has been done by some authors) to make the form of the chloroplast the basis of classification (Carter 1920b, p. 316).

Although some experimental studies have been conducted on chloroplast development in relation to environmental conditions (Kopetzky-Rechtperg 1954 and p. 59), the observations such as those of Carter on *Staurodesmus grandis* (see p. 53) have yet to be followed up under controlled experimental conditions. It would be instructive also, to investigate, by appropriate experiments, the validity of Elfving's hypothesis outlined above (p. 53) about cell size, the axile chloroplast, and light penetration to its core. Teiling (1952) has examined this relationship in a semiquantitative way from data in the Wests' Monograph on British Desmidiaceae and finds a reasonable degree of correlation between absolute cell size expressed as length and chloroplast-pyrenoid type. Thus of the 177 species of *Cosmarium* with axile chloroplasts with a central pyrenoid (monocentric chloroplasts), 65 % do not exceed a length of 40 µm and 92 % do not exceed 50 µm in length. In the genera *Staurastrum* and *Staurodesmus* 76 % of species with this chloroplast type do not exceed 50 µm. Similarly in colonial-filamentous genera *Cosmocladium*, *Oocardium*, *Sphaerozosma*, *Onychonema*, *Spondylosium*, *Hyalotheca* etc., most of which possess short cells of from 20–30 µm long, the chloroplasts are monocentric.

Additional confirmation comes from the genus *Euastrum* in which the monocentric species rarely exceed 70 µm while species with tetracentric chloroplasts (see p. 49) may be greater than 200 µm in length. Also the robust *Xanthidium armatum* (Fig. 38: 4) which has parietal chloroplasts exceeds 200 µm in length while the di- and tetracentric species are between

50–100 μm, and the monocentric less than 40 μm long.

DEVIATIONS IN CHLOROPLAST DEVELOPMENT

In her classic papers on desmid chloroplasts Carter described several abnormalities in chloroplast development. For example with reference to *Euastrum* she states (Carter 1919a) 'It is quite a common thing, to find in many species specimens in which the central axis is much shorter than usual, leaving a space either at the apex or base of the semicell, or both. Occasionally very irregular chloroplasts are found in which the central axis may be perforated or displaced, together with the radiating plates, from its usual position'. In a study of the effects of iron deficiency on *Micrasterias denticulata* in culture, Hygen (1943) reported that the most conspicuous manifestation of such a deficiency was in the chloroplasts, especially in the polar lobes. It is typically seen as a recurving of the chloroplast from its contact point with the wall at the apical notch. In extreme cases it occurs as an axial fissure, dividing the chloroplast into approximately two equal parts. Other aberrations resulted in the appearance of an open space in the central part of the cell, close to the nucleus. Waris and Ronhianen (1970) report chloroplast defects in *M. fimbriata* induced by gamma rays. Shrinkage of chloroplasts and white spotting, as well as dissolution of pyrenoids, have been shown to be common as a consequence of treatment of cells of *M. torreyi* and *M. sol* with phenothiazine derivatives and antihistamines (Simola and Haapala 1970).

Hygen (1943) followed the deviations which he produced through successive cell divisions and found that when the chloroplast is split by the formation of a straight axial fissure, the symmetry of the forces which seem to regulate the streaming movement of the parent cell contents into the developing semicell (see p. 117), are not interfered with. Thus this type of deviation tends to be transferred through successive generations. He also noted that during the division of cells with these deviations, small fragments may become separated from the main body of the chloroplast and assume a more or less globular form. These seem to survive a series of divisions with little change in form or size, and there was no evidence that they ever became again incorporated into the main plastid. Hygen reports also that 'with growing age, they assume an even darker colour, which by examination . . . can be seen to be due to an exceptionally dense grana structure. Very remarkable also is the total lack of stroma starch in these ageing, dark-green plastid bodies, even if pyrenoids may be present, giving the impression that they have lost their photosynthetic power, in spite of an extremely high chlorophyll content'. It should be noted in this context

E

that Pringsheim (1930), found a pronounced stimulation in the formation of ridges in the chloroplasts of *Micrasterias* in response to an adequate supply of iron. A thorough analysis of cell division in desmids with irregular chloroplasts might, as Hygen suggests, be of value in studying the 'distribution forces' active in constituting internal semicell development and morphogenesis (see Chapter 8).

THE FINE STRUCTURE OF THE DESMID CHLOROPLAST

The examination by electron microscopy of the fine structure of a range of algal chloroplasts (Gibbs 1970) has revealed that nearly every algal division has a distinctive chloroplast type, each with a fairly consistent structure. In very general terms all chloroplasts consist of a series of lamellae composed of basic membrane units, *thylakoids*, each of which is a flattened sac composed of a single membrane. The arrangement of the lamellae is, however, quite distinctive in different groups of algae. In most, the chloroplast, in addition to being enclosed in its own double-membraned envelope, is surrounded by a sac of endoplasmic reticulum. However, in the classes Chlorophyceae and Prasinophyceae of the division Chlorophyta the chloroplasts are limited solely by a double-membraned chloroplast envelope. Gibbs (1970) reports that Keirmayer and Fairbanks have communicated to her that in the desmids *Micrasterias denticulata* and *Closterium acerosum* a layer of endoplasmic reticulum is closely associated with the chloroplasts which suggests a fundamental difference in the class Zygnemaphyceae as compared with other green algae. However, she comments that there are numerous gaps in the layer of endoplasmic reticulum adjacent to the chloroplast; also that ribosomes are present in both of its membranes. She therefore concludes that in desmids elements of the rough endoplasmic reticulum of the cytoplasm lie close to the chloroplast but do not form an uninterrupted envelope surrounding it as found in other algae divisions. A further distinctive feature of chloroplasts of the Chlorophyta and hence of the class Zygnemaphyceae, is that starch is stored inside the chloroplast; a higher plant attribute.

The basic structure of desmid chloroplasts seems to be very complex since not only do they contain an abundance of lamellae but these show structural variations during the life of the cell (Chardard 1964, 1975). It has been demonstrated (Chardard and Rouiller 1957, Chardard 1965) from both light microscope and EM studies that in the relatively few desmid chloroplasts examined, structures analogous to the grana of higher plant chloroplasts are apparent, grana being used specifically to refer to a stack of adpressed thylakoids (Gibbs 1970). Thus in their EM examination of chloroplasts of desmids from three different genera, Chardard and Rouillet (1957) concluded that *Micrasterias* had grana very similar to those of higher plants while those of *Cosmarium* and *Closterium* were less well developed and rudimentary. In a later study of these two latter genera Chardard (1965) found grana-like structures in both, but these he termed *pseudograna* to emphasise their resemblance to typical grana and yet to indicate a difference between the organisation of their thylakoids and that of higher plants. It should be noted in this connection that in studying *Carteria* (Chlorophyceae; Volvocales) Lembi and Lang (1961) applied the term 'paseudo-grana' to the piles of thylakoids which they found in its chloroplasts, because these did not contain complete discs but were open at one extremity.

Drawert and Mix (1961) in their EM study of *Micrasterias rotata* have shown that the grana, which they observed were labile structures, appearing only under certain conditions which seem to depend on the metabolic activity of the cell. Under certain unfavourable conditions, as for example in the presence of antibiotics, the grana do form, which suggests that they appear as a consequence of the slowing down of the development of the chloroplast.

The fundamental question still to be answered, however, is whether the grana-like piles of thylakoids observed by various investigators truly correspond to higher plant grana, or are merely a consequence of a secondary multiplication of the thylakoids (Menke 1962, 1965). An approach to this problem has been undertaken by Chardard (1975) by investigating the consequences of darkening and subsequent re-illumination on the submicroscopic structure of the chloroplasts of *Cosmarium lundelli* and *Closterium acerosum*.

After a period of up to 5 months in the dark some of the cells of *Cosmarium lundelli* remained green and the basic lamellar structure of the chloroplasts was still apparent, though less complex in form than those of illuminated chloroplasts. Indeed as a result of darkening, chloroplast development it rapidly slowed down, as in the case of the treatment with antibiotics mentioned above, and there was a considerable simplification of the basic architecture of the lamellae. From these observations, however, Chardard concluded that because the lamellar system persists in the dark it must be a fundamental structure. He also found that the degree of involution of the thylakoids was almost unaffected by darkening.

Under normal conditions of illumination and therefore of cell metabolism, so that the desmids are dividing rapidly, Chardard believes that the chloroplasts must be in a state of continuous reorganisation and hence their development and differentiation must occur simultaneously. He suggests that differentiation of the thylakoids which form the grana is by evagination and multiplication, and by the invagination of their walls. He also observed that they are produced in a rather disorderly fashion instead of as regular accumulations, which is the case in the formation of the grana of higher plant chloroplasts.

The action of reduced light on *Closterium acerosum* provided little or no information about the grana-like structure of its chloroplasts. However, some information was gained about other aspects of thylakoid development—a still incompletely understood phenomenon, not only in algae, but in all plants. It is generally supposed that they are formed by the coalescence of small vacuoles, although these do not become abundant in the normally developing chloroplast. On the contrary, Chardard reports the existence of large vesicles of variable shape and filled with a granular substance. These vesicles are said to elongate and fuse, giving rise to a granular stromatic substance. This process apparently has no counterpart in other plant groups. Chardard reports that in the later stages of development a granular substance is disposed between the bundles of lamellae, but that within the same bundles there is a clear space. He poses the question as to the nature of the clear substance which seems to fill the chloroplast before the development of the stromatic substance.

There is clearly much still to be discovered about the submicroscopic structure of desmid chloroplasts, not only with regard to their basic structure and development, but also under various environmental conditions.

THE CHLOROPLAST AND ASSOCIATED CELL ORGANELLES

1. CHLOROPLAST AND NUCLEUS

In her detailed studies of desmid chloroplasts Carter (1919a) stressed that a very intimate connection exists between chloroplasts and the cell nucleus. This can be seen most clearly where there are two axile chloroplasts in each semicell, as in many *Cosmarium* spp. In these the basal part of each chloroplast is drawn out into a string-like portion which is appropriately attached directly to a corner of the nucleus. These observations have been confirmed by the more recent study of Korn (1969a) whose useful diagramatic representation of the relationship is reproduced in Fig. 27. It is of interest to note that Carter was one of the first to observe the effects of centrifugation on cell inclusions and she notes that after such treatment the nucleus becomes displaced from its normal position in the isthmus and can be seen in one of the semicells 'dragging behind it the chloroplasts by means of their string-like attachments, the connections between them still unbroken'. She concludes that there must therefore be a very vital connection between nucleus and chloroplasts. However, as far as can be ascertained, no one has to date explored the nature or significance of this presumed relationship.

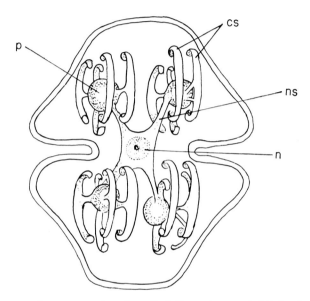

Fig. 27. A diagrammatic representation of a cell of *Cosmarium turpinii* Bréb. showing the four chloroplast units and their connection to the nucleus (n) by strands (ns). Note also the connection between the pyrenoids (p) and the chloroplast strands (cs). (After Korn.)

2. CHLOROPLASTS AND CYTOPLASM

Carter also observed thread-like attachments between the edges of the chloroplast plates and the layer of cytoplasm lining the cell wall. She found this to be particularly the case in cells which possessed chloroplasts insufficiently massive to extend to the cell wall at all points, especially in developing semicells. These are especially noticeable in the saccoderms *Netrium digitus* (Fig. 28) and *N. oblongum*, though they do occur in many placoderms and Carter (1919, p. XX, Fig. 19) depicts a cell of *Micrasterias oscitans* (?) with very prominent wall attachments radiating from the edges of the chloroplasts. Kopetzky-Rechtperg (1954) has shown that seasonally the chloroplasts of *Netrium digitus* show varying degrees of atrophy, and as this condition progresses the thread-like strands become longer and hence somewhat more prominent (Fig. 28, a–c).

3a. PYRENOIDS

In considering chloroplast evolution, attention has already been directed to the differences which occur in pyrenoid numbers and position during the postulated translocation from its axile to parietal position. Thus, in species with axile chloroplasts the pyrenoids are usually found in those parts of the cell where the chloroplast is most massive, so that they may be in the central core or in the peripheral rays. The number of these organelles embedded in the centre of the axile chloroplast would seem to depend on cell length. Hence, in long narrow *Closterium* spp. they occur in an extended row and may be very numerous. In many desmids where the axis is very limited in length there is usually only one pyrenoid in the axis of each chloroplast.

The number of pyrenoids is not fixed and the physiological condition of the cell would seem to determine how many are present. Irregularities in number were first noted by Lütkemüller (1893) and confirmed by the studies of Ducellier (1917) and Carter (1919a). As the last author has pointed out, variations in pyrenoid numbers are not always noticed especially in small biradiate cells. When pyrenoids divide, according to Carter (1919a), usually by budding or by the gradual constriction of one into two, the newly formed pyrenoids separate from each other almost immediately.

For example, some species of *Cosmarium* have two axile chloroplasts in each semicell each with a single pyrenoid. If either of these divides the products cannot separate for the purely physical reason that the chloroplast at this point forms only a very narrow film around them. Thus two pyrenoids remain closely addressed in the position of each original one. Subsequent divisions may result in groups of 3 or more pyrenoids; but because they are so closely associated and have no room to separate, they cannot readily be recognised as multiple pyrenoids. Only in stained preparations can such pyrenoid groups be distinguished.

The conditions which determine the division of pyrenoids are not at all clear and observations of natural populations give no clue as to the possible causal factors. One desmid in a collection may contain a few large pyrenoids, while another of the same species may contain many smaller ones (Carter 1919a). It has even been observed that the extent of pyrenoid division may vary in different parts of the same semicell. Thus Carter instances the case of a specimen of *Xanthidium brebissonii* containing two axile chloroplasts in a semicell where one chloroplast had a single very large pyrenoid and the other had a group of small ones (Fig. 28, d–h).

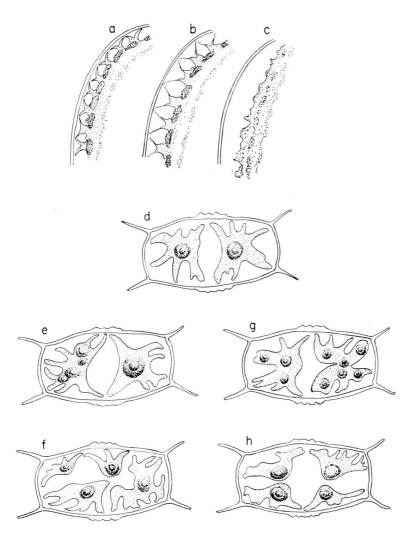

Fig. 28. a–c. Diagram showing the thread-like attachments between the edges of chloroplast plates and the cytoplasm lining the cell wall in *Netrium digitus* (Ehr.) Itz. and Roth. The series a–c indicates the gradual atrophy of the connecting strands as environmental conditions change (after Kopetzky-Rechtperg 1954). d–h. A series showing transverse sections of various individuals of *Xanthidium brebissonii* Ralfs. illustrating the variation in number and size of the pyrenoids and transitions from the axile to the parietal disposition of the chloroplasts (after Carter).

Following the centrifugation of cells of *Closterium moniliferum* for physiological experiments, Weber (1965) noted the development of unusually large numbers of pyrenoids. Normally its chloroplasts contain from 2 to 6 pyrenoids along their median axis. With centrifugation the pyrenoids were readily displaced, and, although this treatment may result in irreparable damage to many cells, in those that survived new pyrenoids were seen to arise 'de novo' within 36 hours following treatment. Centrifuged cells were found to contain an average of 10.7 and a maximum of 18 pyrenoids (5.35 and 9 per chloroplast), whereas controls contained an average of 7.9 and a maximum of 12 (3.45 and 6 per chloroplast). Displaced pyrenoids appeared unchanged with time and were passed on to daughter cells during subsequent divisions, though a marked depletion of pyrenoid (and stroma) starch was noticed in treated cells. This was assumed to be due to the mobilization of carbohydrate reserves during recovery.

Similar studies using centrifugation might provide valuable information about chloroplast–pyrenoid relationships and might profitably be employed in studying the fine structure of pyrenoids during their formation, about which there is an almost total lack of information.

It is not unusual to find numerous darkly stained globules scattered amongst the pyrenoids in the chloroplasts of some desmids. Personal observations suggest that they occur very commonly in the large *Xanthidium armatum*, though Carter (1919a) records their occurrence in *Cosmarium ochthodes*, a *Cylindrocystis* sp., *Micrasterias denticulata* and several other species of *Cosmarium* and *Closterium*. In *C. ochthodes* they are present in the peripheral part of the radiating chloroplast plates, but more often, and especially in other desmids, they occur in the reticulated films of chloroplast which stretch from the latter over the interior of the cell wall (cf. p. 60). Carter (1919a, p. 226) remarks that in stained preparations they appear like small naked pyrenoids and that they react identically to the 'pyreno-crystals' of large pyrenoids. She found that they stain brilliantly with acid fuchsin and thus comments that they are undoubtedly proteinaceous.

As to their origins and formation, Carter suggests that they seem to have no relation to the large pyrenoids and are probably not derived from them, but would seem to be formed 'de novo' when conditions in the cell are favourable for the storage of large quantities of food. However, they still await detailed examination and study.

3b.   PYRENOID FINE STRUCTURE

Electron microscope studies of desmid pyrenoids as entities with a close relationship to chloroplasts (quite apart from investigations of their development) are few. Leyon (1954) examined *Closterium acerosum* and *C. lunula* in this connection and observed, in cells which had been kept in weak light for several months to reduce their starch content, that the interior of the pyrenoid showed a dense structure consisting of tightly packed lamellae. In many sections it could be seen that 3, 4 or even 5 chloroplast lamellae come together to form a single group of adpressed thylakoids within the pyrenoid. As can be seen in the diagrammatic representation of the *Closterium* pyrenoid (Fig. 29) the much thicker thylakoids of the latter, which are thought to contain some additional

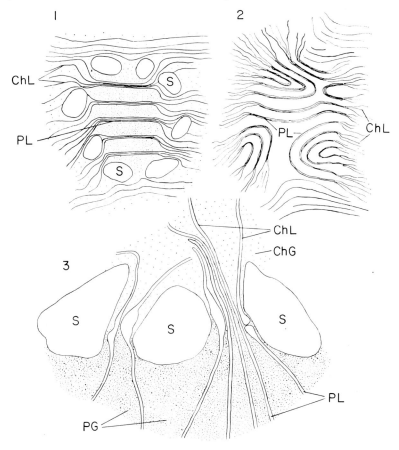

Fig. 29. Diagrams interpreting the submicroscopic structure of pyrenoids in desmids: 1 a simple pyrenoid without starch (after Lyon); 2 pyrenoid of *Closterium acerosum* (Schrank) Ehr. without starch (after Lyon); 3 part of the pyrenoid and chloroplast with starch grains of *Micrasterias rotata* (Grev.) Ralfs. (After Drawert and Mix.) Ch L = chloroplast lamellae; Ch G = chloroplast ground substance; PL = pyrenoid lamellae; PG = pyrenoid ground substance; S = starch grains.

dense material, are grouped in parallel curved series which have been described by Leyon as resembling sections cut from an onion and then have been rearranged with their internal surfaces turned outwards. In their EM examination of *Cosmarium lundelli* Chardard and Rouiller (1961) noted that the bands of the chloroplast lamellae, which are composed of series of adpressed thylakoid discs, are reduced to single discs where they traverse the pyrenoid (Fig. 29).

Following a detailed EM study of the pyrenoids of *Micrasterias rotata* Drawert and Mix (1972) suggest that three elements can be distinguished in these organelles: (1) ground substance, (2) a double lamella system, and (3) a starch envelope. The latter, they point out, can be lacking under certain circumstances, especially after culturing with streptomycin. The ground substance and double lamellae form what they term the 'pyrenophor' (Fig. 29: 3). The double lamellae of the pyrenophors are connected to the lamellae system of the rest of the chloroplast through grana-like bridges but are continuous, as Leyon has indicated (see Fig. 29: 1, 2), with the chloroplast lamellae. Within the pyrenophor the lamellae may vary considerably in number depending on the physiological state of the cell. Also they may be considerably folded and run in various directions (Fig. 29: 2).

An element specific to the pyrenoid would seem to be the dense, markedly osmiophilic ground substance, which is separated from the ground substance of the stroma by the compactness of the pyrenoid itself and by inflations of its double lamellae towards the periphery (Fig. 29: 3).

The pyrenoid starch is not distinguishable from that of the stroma and it occurs only on the periphery of the pyrenophor. Moreover its formation would seem to be limited to this locality within the ground substance and never within the double lamellae. However, it is possible to interpret the occurrence of a reservoir of starch on the inner side of the pyrenophor in tangentially cut EM sections.

Occasionally double pyrenoids have been observed but it is doubtful whether these represent stages in division. However, there appears to be some evidence for a redevelopment of pyrenoids. Under the influence of streptomycin, not only do the starch bodies disappear but also the pyrenoids themselves, under which conditions their ground substance is significantly reduced.

3c.   PYRENOID FUNCTION

Because of their intimate connection with chloroplasts, investigations have been undertaken to discover whether pyrenoids also contain chlorophyll. By simple microdissection Leyon (1954) was able to disrupt the walls of

Closterium acerosum, which had been fixed in 2% formalin for 0.5 h, and extrude parts of the cell contents so that isolated pyrenoids were obtained. Initial microscopic examination showed them to be red (fluorescence?) but following slight cover glass pressure they were transformed into a green smear suggesting the presence of chlorophyll in them.

The presence of chlorophyll in pyrenoid starch has been indicated through light microscopy by Wartenberg (1965) in *Mesotaenium* and interpreted as evidence for the synthesis of chlorophyll in the pyrenoid of this saccoderm desmid. According to this author, the pyrenoid starch represents a vehicle for the transport of chlorophyll, especially chlorophyll 'a' from its site of synthesis in the pyrenoid to the chloroplast.

Chardard (1964), in an ultrastructure study of the saccoderm *Mesotaenium caldariorum* and the placoderm *Penium margaritaceum*, examined their chloroplast structure and described the fine structure of the pyrenoids. With regard to the latter in *M. caldariorum*, he comments that the double lamellae of the pyrenoids are free and thicker than those traversing the chloroplast. The intersaccular space is not uniformly free and is partially constricted. Thus, he comments that the pyrenoid lamellae represent the margins of saccules or tubules. Around the pyrenoid are clear structureless masses, the starch grains, surrounded by a fine border.

The pyrenoid of *P. margaritaceum* is an ovoid structure composed of granular material which appears dense after osmium tetroxide fixation and is traversed by clear canals. However, after potassium permanganate fixation, the ground substance of the pyrenoid, by contrast, appears clear and is traversed by pairs of lamellae continuous with those of the chloroplast.

The close association between pyrenoids and starch production has long been recognised and ultrastructural evidence from numerous algal studies suggests a process of secretion of the precursors of starch synthesis from the pyrenoid. Thus pyrenoids could be regarded as temporary storage regions of the early products of photosynthesis which under conditions conducive to high production could be readily converted into more permanent storage material in the form of the typical starch sheath. As such, it would be expected that they contain much greater concentrations of enzymes involved in carbohydrate metabolism than the rest of the chloroplast.

As part of a study relating to the enzyme distribution in cells parasitized by various phycomycetes, Cook (1967) investigated acid phosphatase and nuclease activity in *Closterium acerosum*. Using the Gemori method of detection of acid phosphatase he found that maximum reaction varied from within the pyrenoid to the central core of plastid between adjacent pyrenoids. Acid nuclease activity was demonstrated in both nucleus and chloroplast, in which it appeared in granular and non-particulate form, the granular tending to be arranged around the periphery of the pyrenoid.

The association of acid phosphatase and nuclease with the pyrenoid suggests to Cook that this special region of the desmid chloroplast may have functions other than starch synthesis. Although its exact role is uncertain, for Cook the possibility exists 'that these enzymes could play a part in the anabolic processes within the highly structured membrance system of a plastid.'

Because, as indicated above, it seems that pyrenoids can, without great difficulty, be removed from desmid cells many of which are large enough to be subjected to microdissection. Such algae, the majority of which can also be easily cultured under a fairly wide range of physical and nutritional conditions, could provide a most valuable source of material for much needed biochemical and physiological studies of these still enigmatic organelles.

3d. CHLOROPLAST MOVEMENT

The chloroplasts of many plants possess the ability to orientate themselves into such a position within the cell that the light falling onto them approaches the optimal for photosynthesis. These movements may be either to avoid high light intensities (high-intensity responses)—also UV light is intensely damaging to living matter, including chlorophyll—or when intensities are low, the maximum area of chloroplast is turned directly towards the incident light (low-intensity response).

Placoderm desmids, many of which, however, are motile (see p. 67), do not possess this faculty and the whole cell, instead of merely the chloroplast, can orientate itself or move into or away from the existing light conditions. As in other algae and higher plants the chloroplasts of some saccoderms are capable of light-intensity responses. As pointed out by Muhlethaler to bring about light-dependent chloroplast orientation, the cell must not only absorb light, but also have the ability to determine light direction and distinguish between intensities. Also it must possess a cytoplasmic system that physically moves the chloroplast, and this must be supplied with energy from the cell's metabolism, since energy received from the light stimulus may not always be sufficient for chloroplast displacement. Of very special interest is the nature of the photoreception absorbing the light necessary to induce chloroplast movement. This aspect of the problem has been studied in low-light intensities in the saccoderm desmid *Mesotaenium caldariorum* by Dorscheid and Wartenberg (1966). Their results are particularly intriguing since most reports, relevant to this phenomenon with other algae and higher plants, indicate action spectra for chloroplast movements which differ from chlorophyll absorption curves (Haupt 1960). In *Mesotaenium*, however, action spectra show the initiation of such movement to be controlled by light absorbed by chlorophylls 'a' and 'b', whereas in a second phase of chloroplast movement the phytochrome system would seem to be involved in addition to the chlorophylls.

3e. PHOTOTAXIS IN DESMIDS

Although some saccoderms seem to exhibit little motility, species of *Cylindrocystis* and *Netrium* have been observed to move actively, especially in freshly collected material (personal observations). However, a detailed study of such activity has yet to be carried out. Hoshaw and Hilton (1966), in studying the sexual cycle of *Spirotaenia condensata*, comment that the deposition of mucilaginous material from one end of these cells imparts motility and this results in the formation of long, often coiled mucilaginous tubes (see their Fig. 1).

In the first monograph to be written on desmids, Ralfs (1848) described how many placoderm species moved to the surface of the mud when placed in a vessel and noted a particularly marked response with *Tetmemorus granulatus* and *Penium brebissonii*. He presumed this response to be stimulated by light 'rather than to any voluntary effort'. Braun (1851) also noticed that light-induced movements were especially marked in *Penium curtum*, the cells of which not only turned to align their longitudinal axes in the direction of the light, but also in any given cell it was the younger (daughter) semicell that always pointed towards the source. In 1880 there were reports on desmid phototaxis by Gobel and Stahl, the former with reference to *Micrasterias*, the latter *Closterium moniliferum*. Gobel reported that the maximum surface area of *Micrasterias* always orientated towards the direction of the light beam, though Bendix (1960) found no light conditions that would consistently induce this position. Indeed with a stimulating light beam entering a vessel containing *Micrasterias* at a 'grazing angle' to the bottom and at a range of intensities from below the threshold of positive phototaxis to intensities even inhibiting negative phototaxis, the cells generally remained flat on the bottom. Stahl, investigating *Closterium*, found that although the cells lined up in the plane of the incoming light the younger semicell did not, as Braun noted, point towards it.

Stahl (1880) observed that although the lunate cells of *Closterium* could glide along the bottom of a vessel in response to light, their most frequent means of locomotion was an end-over-end flipping. Bendix

(1960) recorded similar types of movement in *Micrasterias* though of the two, flipping was much less frequent. The connection between desmid movement and the localized secretion of gelatinous material through the pores (see p. 87) was first suggested by Klebs (1885), who, in addition, observed that *Pleurotaenium* orientated in the direction of a light source at low intensities, but perpendicular to it at high intensities. The mucilage is usually secreted through large pores situated at, or towards, the apices of the cells, and it has been observed in some species that irregular, worm-shaped masses of considerable length may be secreted in less than an hour. These secretions can be readily demonstrated in species of *Closterium* by placing the cells in dilute Indian ink (Fig. 30). Their somersaults, each of which can occur in as short a time as 6 minutes, depending on temperature, can be easily observed.

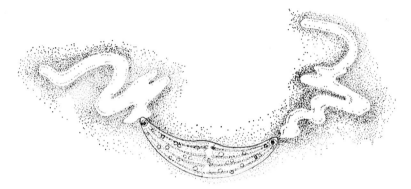

Fig. 30. Phototactic movement in *Closterium leibleinii* Kütz. caused by the secretion of mucilage from terminal pores. As this illustration shows, the secretions become much more apparent when the cells are placed in a solution of Indian ink.

Few detailed studies have been undertaken to investigate the physiology of phototactic responses in desmids. Bendix (1960) noted that *Micrasterias rotata* loses most of its phototactic response shortly before cell division and this is not regained until the reorganisation of the daughter cells is visibly judged to be complete.

In other studies with *Micrasterias rotata*, which moves at a maximum of 3 mm/hour, Bendix demonstrated that the cells showed two separate phototactic responses when placed in a light beam dispersed by a large prism. One response was in the place of the beam (PN response), the other perpendicular to it (L response). The latter response was unidirectional and caused cells to move out of the long wavelength end of the spectrum but no further than 570 μm. This response was independent of the direction of the energy gradient across the region in which it was observed and was not correlated with the chlorophyll absorption peak in red light (665 μm). Also it took place in a range of intensities from those too low to support division to high enough to irreversibly inhibit it.

The PN response was quite different, for as light intensity was increased it appeared to show two peaks. Thus it went to a positive maximum, back to zero, and a small null response, and then to a negative maximum and again to a zero response. Bendix found the threshold of the positive PN response in white light was slightly below 2/0.5 lx, while in monochromatic light at 690 μm it was only 5×10^{-4} μw/cm² (1.5×10^5 quanta/s/cell). This positive response saturated in white light between $18.6–55.7 \times 10^{-2}$ lx, in monochromatic light at 690 μm it was 5×10^{-3} μm/cm². The action spectrum for the response obtained indicated that chlorophyll 'a' and possibly another unidentified chlorophyll act as photoreceptors. Taylor and Bonner (1967) have implicated a phytochrome as the receptor in the saccoderm *Mesotaenium*.

A somewhat superficial study of motility in *Closterium acerosum* by Yeh and Gibor (1970) reported a lack of a clear relationship between the direction of movement by this desmid and the direction of a light source. Confirmation of the fact that photosynthetic activity is a prerequisite for movement is suggested by their experiment in which 10^{-5}M dichloro-phenyl-dimethylurea, a photosynthesis inhibitor, reduced trail production from 27 to 7% of illuminated cells.

Häder and Wenderoth (1977) have shown, by light-trap experiments using species of *Cosmarium*, *Micrasterias* and *Euastrum*, that three light-induced motor responses, known in other microorganisms, will cause photoaccumulations. During the initial phases, the desmids orientate phototactically so that their long axes become aligned parallel to the incident light; then they advance towards it. Photokinetic reactions, though indicated, play only a minor role, while photophobic reactions, the third type, have been demonstrated for the first time. These involve either a reversal of movement, or a sideways swing in a different direction. They suggest that in the perception of light direction a mechanism is involved which simultaneously assesses intensity at two or more sites within the cell. Also phototactic reactions revealed a marked circadian rhythm, and in long-term population experiments, phases of activity were remarkably synchronous, so that rings of movement could be seen in their culture dishes.

THE APICAL VACUOLES IN THE ARCHIDESMIDIINAE

Several investigators have been greatly intrigued by the apical vacuoles that occur at each end of the cells of desmids belonging to the sub-order

Archidesmidiinae, such as the genera *Penium, Pleurotaenium* and especially *Closterium* (Fig. 31). Because these vacuoles contain one or more clearly visible crystals in a state of constant agitation, and because cells with these organelles are the most motile of all desmids, the inference has been drawn that these striking structures perform the function of statoliths (Steinecke 1926). Frey (1926) does not believe that this is their function but rather that they play some role in osmoregulation.

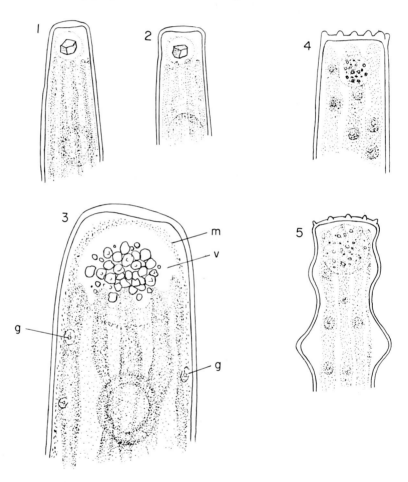

Fig. 31. Terminal vacuoles and their included granules in: 1 and 2 *Closterium angustatum* Kütz.; 3 *C. didymotocum Corda;* 4 *Pleurotaenium ehrenbergii* (Bréb.) de Bary; 5 *P. nodosum* (Bail.) Lund. Note that in 3, small granules (mitochondria 'm') surround the vacuole and these, like those in the rest of the cytoplasm, are in a state of constant agitation as they circulate round the vacuole. Isolated large granules (of BaSO$_4$ 'g') occur in the cytoplasm and in whose stream they are moved.

There are two categories of apical vacuoles; the most common being the spherical vacuoles such as are found in *C. acerosum, C. didymotocum* and *C. ehrenbergii* and in some species of *Penium* and *Pleurotaenium* (Fig. 31). Less common are the elongated, cone-shaped type found, for example, in *C. dianae* and *C. parvulum*, which extend from the end of the chloroplast to the tip of the cell (Fig. 32). They have been investigated by Leblond (1928), Chadefaud (1936) and in considerable detail by Kopetzky-Recht-perg (1931) and Wurtz (1942), all of whom remark on the instability of the conical vacuoles in that they show continuous changes in shape and may even fragment (Fig. 32).

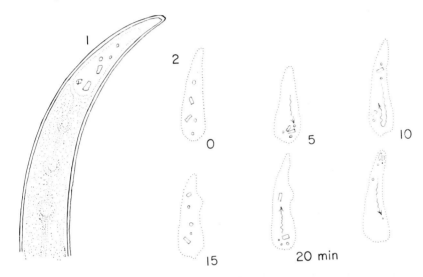

Fig. 32. 1 Apical cone-shaped terminal vacuole in *Closterium dianae* Ehr. illustrating the scattered, moving granules. 2 Movements of apical granules in a vacuole of *C. dianae* at 5-min intervals (after Wurtz). The lower right diagram of the series shows the position of the granules in the tip of the cell when it was moved from a horizontal to vertical position on the microscope stage.

In cells with spherical vacuoles there may be only one large crystal contained within each, but other spherical vacuoles may contain few or as many as 100 crystals. These may be agglomerated into a centrally placed, jostling, moving mass in which the individual crystals are difficult to count. The elongated conical vacuoles always contain numerous crystals which move about the vacuoles (see below and Fig. 32). Desmid taxonomists have suggested that the number of crystals is constant and characteristic for each species. Whilst this may apply to some, Laporte (1931) doubts whether this is always true and Kopetzky-Rechtperg (1931, 1933) suggests that the number present depends on the physiological state

F

of the cell, and especially on its age. Significant in this connection are observations made by Wurtz (1942) on the crystals in the elongate vacuole of *C. dianae* in which the number of crystals, which are of different sizes, varies from 5–15. In a cell which had 5 crystals, 2 large and 3 small, he found that by placing the microscope with which he was examining this cell in a horizontal position for a few moments and then returning it to the vertical he augmented this number by 4 new crystals, thus increasing the total to 9. He inferred that crystals outside the vacuole had been carried into it presumably by the change in the direction of the force of gravity (Fig. 32). Crystals can also leave the vacuole.

In surveying the literature relating to the nature of *Closterium* crystals, all references to them suggest that they are of calcium sulphate. Fischer's (1884) attempts to show this were, however, based on negative reactions, and Kopetzky-Rechtperg's (1931) more detailed analysis is not completely convincing. Because of these doubts, *Closterium* cells have recently been examined with a JOEL 100S EM with a scanning and 51EA attachment and subjected to X-ray analysis with a LINK energy dispersion system. The first specimens to be analysed were of *C. ehrenbergii* from the highly calcareous River Lambourn, near Newbury, Berkshire. Freshly collected individual cells were micro-pipetted onto both aluminium and copper stubs and subjected to 'cold ashing' for from 5–10 minutes. Numerous crystals could then be clearly seen in the scanning EM at the ends of the cells. More than 50 were, for example, counted in one vacuole of *C. ehrenbergii*. X-ray analysis revealed only small peaks of calcium, but strong peaks indicating barium and sulphur (Brook, Fotheringham, Bradly and Jenkins, 1980).

Following this finding, experiments using cultures of *C. littorale*, *C. ehrenbergii* and *C. moniliferum*, from the Cambridge Culture Collection, were grown in ASM liquid media, and although during the exponential phase of cell growth the crystals tended to disappear, those still found to be present showed characteristic barium and sulphur peaks on X-ray analysis. Cultures of *C. leibleinii* obtained from Cambridge would not grow successfully in ASM medium, but individuals examined from a culture of the desmid grown in soil/water cultures of Kettering loam also contained crystals which produced positive barium and sulphur peaks. *C. striolatum* from a pond in Leicestershire, and *C. intermedium* from an English Lake District tarn, have also revealed barium and sulphur as the major elements in these crystals. Water analyses performed by the Water Research Centre, Medmenham, Buckinghamshire, indicate that barium concentrations in a variety of UK lakes and rivers range between 2–33 µg/litre. Calcium concentrations in contrast are at least 4 orders of magnitude greater.

The presence of very conspicuous crystals in the cells of such species as *C. intermedium* and *C. didymotocum* (also in *Pleurotaenium undulatum*) from waters of vanishingly low calcium content, such as the markedly acidic waters of many Scottish Highland lochans and bog pools, was very puzzling when, until this recent discovery as to their true nature, it was believed that the crystal were of $CaSO_4$. However, it is now less so, since it has been pointed out (Bowen, pers. comm.) that barium is much more abundant in waters in granitic regions than in those in the catchment of sedimentary rocks very much richer in calcium.

The first to suggest that some freshwater algae might accumulate barium was Nicolai who observed in dried filaments of *Spirogyra* X-ray interferences resembling those of $BaSO_4$ (Kreger 1957). Later Kreger and Boeré (1969) showed by chemical analysis that the $BaSO_4$ in *Spirogyra* amounted to 0.4–0.7% of the dry weight and calculated there is 110–185 times as much per unit volume in the living cell as could possibly be present in solution in the environment. Nicolai thought that possibly the $BaSO_4$ was a wall component, but Kreger and Boeré (1969) by electron microscopy have shown it to be present in the cytoplasm and not the wall. The *Spirogyra* crystallites are near, or just above, the resolution of light microscopy, but when examined under a polarizing microscope many very small, highly birefringent particles, showing Brownian movement, were seen. They can also be observed in phase contrast. It may be significant that the crystallites were concentrated in the lower parts of the cells and that they gradually reassumed this position after the filaments were rotated (statoliths?). This observation indicates that the crystals are free in the cell and not connected with any organelle system. Schroter, Lauchli and Sievers (1975) have identified the crystals in the so-called statolith system of the rhizoids of *Chara fragilis* as $BaSO_4$ and their X-ray microanalysis spectra are very similar to those obtained for the range of *Closterium* species examined by Brook *et al.* (1980). Crystals of $BaSO_4$, free in the cells of *Micrasterias rotata*, also a strongly motile desmid (see p. 67), have also been identified by us.

Cultures of *C. littorale* grown in ASM media and virtually devoid of crystals when placed in fresh ASM containing a range of barium chloride to produce from 0.10–10.00 ppm Ba, showed crystal formation at a rate dependent on barium concentration. Thus, at the highest concentration, crystals were clearly evident in all the vacuoles examined within 24 h. Similar crystal-free cells of *C. littorale* were inoculated into Millipore-filtered water from Whiteknights Lake (alkalinity 100–150 mg/l $CaCO_3$) of Reading University. Crystals were observed to have formed in the cells within 5 days. These were found on X-ray analysis to contain not only barium, but also gave positive indications of the presence of strontium.

Subsequent experiments in mineral media containing varying concentrations of strontium alone and with barium, and also barium and calcium have, however, revealed no indication that any of the crystals in these cells contained this rare element, though all again showed barium and sulphur to be their major constituents.

Most observations on the intriguing problem of vacuolar crystal movements have been made on *Closterium* species with spherical vacuoles. Their continual activity is well documented and as a result of careful analysis it became clear that although one component is purely physical and determined by Brownian movement, superimposed on these are other, more irregular motions. Over a hundred years ago, Schumann (1875) suggested that, at least in part, the latter was a consequence of perivacuolar activity of the cytoplasm. He supported this view by stressing that the crystals are not scattered throughout the vacuole but aggregate as a single mass in its centre. Cytoplasmic currents can be readily seen by virtue of the granules entrapped in the cytoplasm which circulate, especially round spherical vacuoles (Fig. 31). They not only change the shape of the vacuole, but the pressures developed by them are transmitted to the crystals inside. These small shock waves, though unequal, are said to be felt in a centripetal direction from all sides. Because of their inequality they cause the crystals to exhibit their characteristic agitation, and because they originate on all sides, the centre of vacuole is clearly the natural place for their aggregation.

Fischer (1883) also considered that it was cytoplasmic movement that caused crystal movement and demonstrated a correlation between crystal agitation and cytoplasmic activity. By adding ammonia, which arrests cytoplasmic movement without causing plasmolysis, to *Closterium* cells, Fisher noticed that the movement of the crystals became less intense, and indeed appeared to show only Brownian movement. Wurtz (1942) carried out similar observations but killed the cells with formalin or alcohol. Cytoplasmic activity then stopped immediately but the Brownian movement of the crystals, observed by Fischer was still visible. Even when cytoplasm and chloroplasts had contracted the crystals still exhibited activity and it was only when the vacuolar fluid increased in viscosity following this drastic treatment that all movement stopped. That crystal activity has a component other than purely physical Brownian movement would seem to have been confirmed by Frey (1926) by simple observations on spherical vacuoles. By inclining his microscope from the vertical to the horizontal position, he noticed that by the resulting sudden change in the direction of gravity, the crystals accumulated at the base of the vacuole but from this mass, occasional crystals were projected from out of the heap. This author concludes that 'ce sont plutôt des courants centripètes, sans doubte des jets d'excrétion qui se superposent au movement brownien'.

Wurtz has followed in detail crystal movement in the elongated conical vacuoles of *C. dianae* and finds that in the basal region, crystal activity is constant, though jerky, and attributable to Brownian movement and does not extend beyond the middle of the vacuole. Suddenly, however, one or two crystals appear to travel by a slow and even movement along the length of the vacuole before resuming their Brownian movement, though he comments that this had nothing in common with the projections described by Frey since the displacement was neither as slow, nor as regular. Wurtz carefully recorded the position of individual crystals at regular intervals and in the case where there were 2 large and 3 small crystals, it was the small crystals which were entrained most easily in these slow, regular but intermittent movements (Fig. 32). With these he also recorded a rolling rotation, though this did not seem to be complete or regular. He also noted that, in accordance with Einstein's law, the Brownian movement of the small crystals was more intense than that of the large ones.

It has already been mentioned (p. 72) that Wurtz repeated with *C. dianae* Frey's experiments in which the microscope was placed horizontally instead of vertically. The crystals fell to the bottom of the vacuole, but in accordance with Stokes' law, the large granules fell more rapidly than the small ones. Indeed, the latter took up to 3 min to reach the bottom. Some, however, adhered to the vacuole wall and fell much more slowly than if following Stokes' law. Also, though most of the crystals remained as group after this radical change in cell orientation, on the base of the vacuole, the small ones still showed the entrainment movements already described. Wurtz invokes Schumann's (1875) explanation of cytoplasmic activity for such movements of the vacuolar contents.

Wurtz has described the changes in the shape of the conical vacuoles of *C. dianae* already alluded to (p. 71). The basal part of the vacuole does not change shape, but he records changes in the shape of the lateral margins due to undulations caused by the cyclosis of the cytoplasm around the vacuole which occur in successive waves. This activity causes marked changes in the pointed upper part of the vacuole which changes shape endlessly. Sometimes the vacuole may fragment and then merge, become markedly pointed at its extremity; also small diverticulae may be produced from the side, producing small independent vacuoles. These then fuse producing again in a few monents an entire vacuole. All such changes are rapid, thus giving an impression of great instability, and appear to be attributable to the actions of the surrounding cytoplasm (see also Leblond 1928).

OTHER CELL VACUOLES

In the introduction to their five volume work 'The British Desmidiaceae'

West and West (1904, p. 5) state that 'the cell protoplasm occupies a large proportion of the interior of the wall and a portion of it always completely lines the inner surface of the cell wall.' The protoplasm contains numerous granules, largely of nutritive nature, and it exhibits a well-marked 'circulatory movement'. Personal experience indicates that especially in certain large species of *Closterium* (e.g. *C. didymotocum*) these movements may be more vigorous than in any other plant cells normally used to demonstrate cyclosis to laboratory classes. Very small granules (mitochondria?) can be seen in these cytoplasmic streams which also surround the apical vacuoles in the genera *Closterium* and *Pleurotaenium* (Fig. 31). However, also carried in the stream may be granules of the size and angular shape of the crystals enclosed within the terminal granules (Fig. 33). (See also p. 70.)

The Wests also comment later that 'when desmids are kept alive for

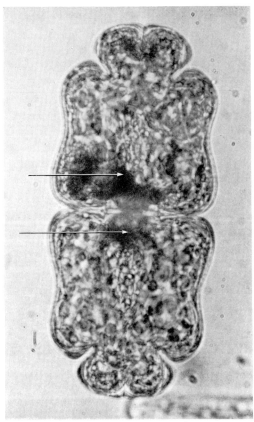

Fig. 33. A photomicrograph of a freshly gathered specimen of *Euastrum crassum* (Bréb.) Kütz. showing (arrows) dark patches of swarming granular material surrounding cell vacuoles.

some time, especially under abnormal circumstances, the protoplasm develops numerous vacuoles, often of large size, all of which become filled with a dense, swarming mass of granular material'. Observations of living material from Scottish lochs has revealed swarming masses of granules to be present even in material examined within a few hours of collection. It has also been found that they are present much more commonly in large than small desmid cells. Moreover, it is questioned whether the violently jiggling granules occur, as the Wests state, within the vacuoles but instead occur within the cytoplasm surrounding the vacuoles. For example, recently examined material from the metaphyton associated with *Sphagnum subsecundum* in the Dubh Loch, Loch Lomond, Scotland, has revealed that, particularly in cells of *Euastrum crassum* and *E. ansatum* (Figs 33 and 34), also of *Xanthidium armatum* (Fig. 38: 4), these granules are so dense in certain parts of the cell that when examined with a ×40 objective they appear as brownish patches. Higher magnification reveals the intense activity described by the Wests, but as shown in the accompanying illustrations, and especially in the side view of *E. ansatum* (Fig. 34), the granules are clearly exterior to the vacuoles. Masses of swarming granules have been observed in cells of *Pleurotaenium ehrenbergii* and these also appeared as brown patches under low magnification (Fig. 34). It should be noted that the position of the granules and their abundance differ in adjacent semicells and some cells (for example *Penium cucurbitum* var. *scoticum*) an aggregation of small swarming granules was observed in only one semicell.

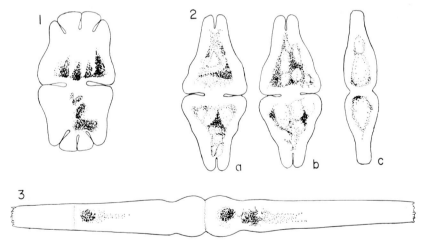

Fig. 34. Line drawings from living material of *Euastrum crassum* (Bréb.) Kütz. (1) *E. ansatum* Ralfs (2 a–c) and *Pleurotaenium ehrenbergii* (Bréb.) de Bary (3) showing the position of granular swarms (mitochondria?) surrounding the cytoplasmic vacuoles.

CHAPTER 4
THE CELL WALL

LIGHT MICROSCOPIC AND MICROCHEMICAL STUDIES

It has been recognised for a considerable time that the walls of desmids are multi-layered (Lütkemüller 1902). Thus the saccoderms whose walls are without pores were originally thought to have walls composed only of two layers, an inner homogeneous layer of cellulose and an outer gelatinous layer of pectose. In some species of *Cylindrocystis* and also of *Mesotaenium* where groups of cells may lie close together as a result of successive divisions, the outer layers of the individual cells may become confluent and so produce extensive mucilaginous masses.

It has been shown that the saccoderm *Netrium digitus* excretes polysaccharides into nutrient medium in culture (Kattner, Lorch and Weber 1977), and thin layer chromatography (TLC) has shown the main components to be xylose, galactose, glucose, galacturonic and glucuronic acids. Hydrolysis of the cell wall showed it to be composed of the same components as the extracellular polysaccharides.

As a result of early studies, all placoderm desmids were correctly described as possessing two layers internal to the exterior gelatinous sheath and differing from one another in their chemical composition. Also there was early recognition that all were perforated by a more or less complex pore system. The innermost, structureless wall layer was found to be of cellulose, the layer external to it (the middle layer) of cellulose impregnated with pectin. It was also observed (van Wisselingh 1912), especially in species of *Closterium*, but also in *Penium*, that iron salts which give many of these desmids a characteristic yellow or rusty colour, were localised in the middle layer.

Carter (1919) found that the cell wall of *Cosmarium biretum* has a particular affinity for the stain iron-haematoxylin so that it became black, entirely obscuring the cell contents. The same phenomenon was noticed in species of *Closterium* and *Penium* and Carter attributed this to the presence of iron salts in the wall which helped to fix the stain. However, the same was not true of *Cosmarium biretum*, for tests for iron gave no positive reaction. In the latter she presumed that the presence of other minerals, probably lime, allowed for the deeply staining properties of the walls.

By means of repeated sonication, and filtration through bolting silk, Lorch and Weber (1972) obtained a mass of cell walls of *Pleurotaenium trabecula* var. *rectum* sufficient for chemical analysis. The purified walls were fractionated by treatment with alkali, dilute HCl, concentrated H_2SO_4, or enzymatically. Thin layer chromatography revealed that the principal sugars present were glucose and galactose, while arabinose, xylose, gluconic and galacturonic acids were less abundant and only traces of rhamnose were found. The enveloping mucilaginous sheaths, which could be removed and analysed separately, have been found to possess, in addition to these cell wall sugars, three unidentified ones. The secondary wall of *Pleurotaenium* was found to react strongly with Schiff's reagent without prior hydrolysis as a consequence of the presence of free aldehyde groups in this wall; but not in the primary wall. Micro-Kjeldahl determinations indicated a total-N content of 0.32% of the cell wall weight, while the lipid content was 1.7%.

Some preliminary histochemical tests and autoradiographic labelling have indicated that the primary wall of placoderm desmids, at least in *Micrasterias rotata*, consists of three distinct fractions (Lacalli 1974b). The first stains with crystal violet, is readily soluble in alkali and may be similar to the mucilage present in secondary wall pores. The second fraction is alcian blue positive, not alkali soluble and this is considered to be a hemicellulose rich in uronic rather than sulphonic acids. These two fractions are considered possibly to constitute the amorphous matrix described by Mix (1966, 1975) but which nevertheless contains a basic network of microfibrils. The third fraction, which is PAS-positive (Jensen 1962) and rapidly incorporates a radiolabel from externally supplied glucose, is also considered to have a microfibrillar structure.

EM STUDIES

EM investigation has revealed that not only placoderm desmids, but all members of the class Zygnemaphyceae possess walls with 3 distinct layers (Mix 1972). These are primary (Gerrath 1975) and secondary walls, fibrillar in their submicroscopic structure and surrounded by an outer amorphous layer. However, it must be emphasised that many desmids of the Cosmariae shed the primary wall following the initial expansion of the daughter semicell following cell division (Fig. 35 and see p. 117). All other members of the Zygnemaphyceae retain the primary wall throughout their lives.

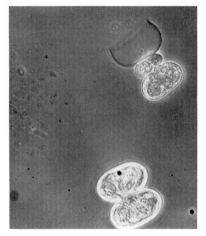
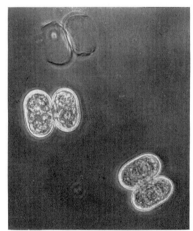

1

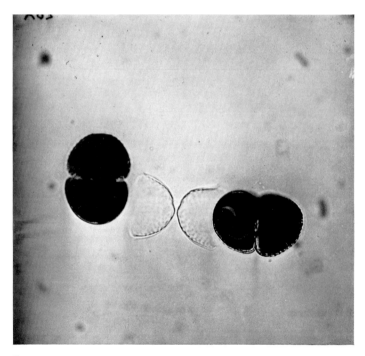

2

Fig. 35. 1 Cast-off primary walls following cell division of *Cosmarium contractum* Kirchn. (phase contrast); 2 Cast-off primary wall of *Cosmarium botrytis* Ehr. Note sculpturing on primary wall.

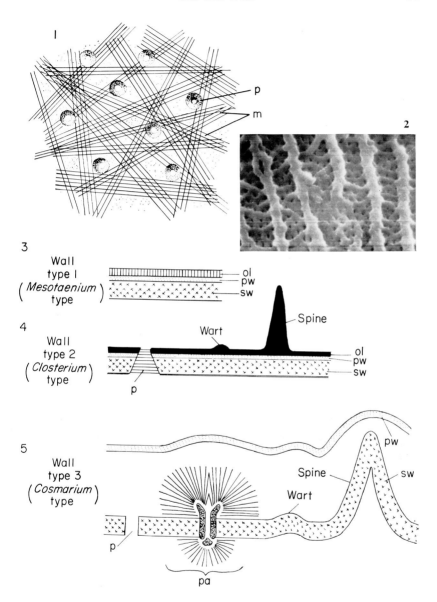

Fig. 36. The Desmid wall. 1 Diagram based on electron-micrographs showing *Cosmarium*-type wall (Type 3) with ribbon-like bands of microfibrils (m) and pores. 2 Scanning EM photograph showing porous mesh-like structure of a portion of cell wall of *Penium spirostriolatum* Barker. 3–5 Schematic diagrams of the three major desmid wall types (for explanation see text). ol = outer layer; pw = primary wall; sw = secondary wall; pa = pore apparatus.

In all desmid walls examined with the EM the primary layer has been found to consist of a network of irregularly arranged microfibrils, though in some elongated forms, such as species of *Pleurotaenium*, there is a tendency towards their longitudinal orientation.

The secondary layer, which develops externally to the primary one, is composed of groups of 8 to 10 parallel microfibrils that form flat, ribbon-like assemblages. These multifibred ribbons occur in layers, lying one over the other and disposed at different angles to one another (Fig. 36: 1). They consist of fibrillar cellulose. This type of secondary wall structure is particularly interesting since it does not appear to occur elsewhere in the plant kingdom (Mix 1975). Kies (1970) has found that a similar ribbon-like construction occurs in the secondary exospore of the zygospore wall of several desmids.

Although the primary and secondary walls of all desmids are fundamentally of the same construction, the outer mucilaginous layer differs in form and composition between species, genera and families, and indeed it determines the surface texture of the wall. This layer consists either of a mucous substance, more or less permeable to electrons, or an amorphous substance largely impervious to them. It is the product of a secretory process initiated after the primary and secondary walls are formed (and shed in the case of the primary in some species), and may continue for an extended period during the life of the cell. Consequently differences may be apparent between old and young cells, between adjacent semicells, or between cell segments where, as for example in the Closteriaceae, there may be intercalary growth (Figs 37 and 45).

The outer layer surface may be smooth or elaborately sculptured. The sculpturing may occur either as longitudinally disposed ridges or as warts or spines which may originate in different layers of the wall. As already indicated, smooth walls occur in the main in the saccoderm desmids, though they are also found in species of *Closterium*, *Cosmarium*, *Staurastrum* and *Staurodesmus*.

Stereoscanning EM studies of desmids were first attempted by Leyon (1969) who explored the external morphology of the walls of *Cosmarium botrytis*. Particularly striking in this paper is a photomicrograph of the isthmal region of the cell which shows very clearly the juncture between adjoining semicells, and how the pores extend into this region. Other species have been examined by Pickett-Heaps (1973) and both studies indicate the potential value of this tool for developmental studies and taxonomy. With reference to the taxonomy of the large and complex genus *Staurastrum* it has been emphasised, as a result of light microscope studies, that the disposition of the often very small apical granules may be a critical character in the separation of the species groups within this genus (Brook 1959b).

Pickett-Heaps' scanning EM micrographs of various *Staurastrum* species confirm the regular and consistent distribution of these potential taxonomic indicators.

Lott, Harris and Turner (1972) have used not only conventional, thin sectioning EM techniques, but also scanning EM and freeze-etching to study the cell walls of *Cosmarium botrytis*. Their freeze-etched pictures show very clearly, as the wall has been shattered, the overlapping, but randomly orientated ribbons of microfibrils described by Mix (1975). Through the examination of freeze-etched walls the interrelationships of microfibrils as well as wall sculpturing and pores can be discerned. Small

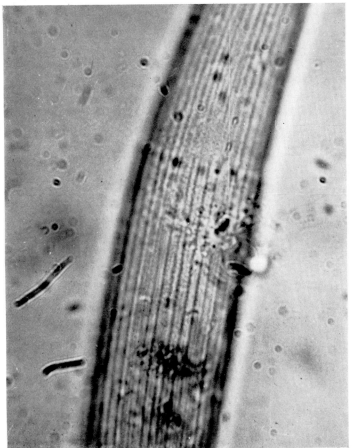

Fig. 37. Part of the empty cell of a *Closterium lineatum* Ehr. showing the longitudinal ridges on the cell wall. Note the difference between the old and new wall, and the well-marked suture line between them (see also p. 104 and Fig. 45 for explanation).

particles embedded in the surface of the wall layers have also been observed in some fractures of the *Cosmarium* wall. These are somewhat similar to those seen in the green algae *Ankistrodesmus braunii* (Mayer 1969) and *Chlorella pyrenoidosa* (Sassen *et al.* 1969) which are thought to be involved in the formation of microfibrils.

The walls of *Micrasterias denticulata*, have also been studied by freeze-etching by Kiermayer and Staehelin (1972) and their excellent scanning EM pictures show very clearly the randomly orientated, characteristic ribbons of secondary wall microfibrils. They have indicated that 'elementary fibrils' of 4–5 nm diam. make up the microfibrils, which are 20–30 nm diam. Two to 15 of these aggregate to form the previously described ribbons. They have also shown that while the microfibrils in the inner layers of the secondary wall would seem to circumvent the pores, those of the outermost layer seem frequently to be disrupted by the pore apparatus. As a consequence of this observation they suggest that the principal component of the pore channel is formed during secondary wall formation. Kiermayer and Staehelin in the same study have also examined the fine structure of the plasma membrane and report that the cleared face following freeze-etching carries numerous randomly distributed particles and there are prominent depressions in it under each pore. Their micrographs show that in the rim of these depressions there are small holes and these they interpret as points of fusion between slime secreting vesicles and the plasma membrane.

The electron microscope has not added greatly to our knowledge of the outer surfaces over that revealed by light microscopy. Thus with the light microscope several types of longitudinal ridges have been found in different species of *Closterium* (Fig. 37), some additional types have been seen with the EM. At this much greater degree of magnification the ridges can be seen to be superimposed upon the primary wall and are formed from an indeterminate, electron-impervious, amorphous substance. Such studies indicate that the arrangement, height and width of the ridges would seem to be characteristic for each *Closterium* species.

Somewhat similar, longitudinally orientated ridges are also evident under the light microscope in some species of *Penium*, as for example *P. spirostriolatum* (Fig. 36: 2), though EM examination of other species has shown that they are by no means as ubiquitous in their occurrence in this genus as they are in *Closterium*. Characteristic, however, of all *Penium* species so far studied submicroscopically is a mesh-like structure consisting of an amorphous outer-layer substance which is electron impervious (Mix 1967, Gerrath 1969) (Fig. 36: 2). The strip-like ridges of *P. striolatum*, warts of *P. margaritaceum* and *P. cylindrus* and spines of *P. spinulosum*, that are visible even by light microscopy, are structures characteristic of these species and have been shown to originate from the same material as the mesh-like structure apparently common to all members of the genus *Penium*. It is of interest to note that as a result of an EM study of the wall of the desmid until then called *Pleurotaenium spinulosum*, Gerrath (1969) was able to show that it was in fact a *Penium* and so has amended its name accordingly.

By light microscopy all species of *Gonatozygon* can be seen to have a wall covered with small dot-like granules or spines. EM examination has shown that these too are of a more or less amorphous nature, impervious to electrons and originating in a layer exterior to the primary wall (Mix 1975). The age of the cell would seem to determine whether the ornamentation is granular (or wart-like) or appears as clearly developed spines. Older cells of *G. brebissonii* have been found to develop a mesh-like structure between the spines comparable to the structures seen in *Penium* species.

As stated elsewhere (p. 79), the majority of the Cosmariae shed their primary wall once the semicell has developed to its full size. EM studies of the developing secondary wall, which forms beneath, show it to be considerably developed before this occurs, and on its surface it is possible to discern the criss-crossing fibrillar ribbons, already mentioned, lying between the pores. The latter are either evenly distributed over the wall

Table 3. Cell Wall Characteristics of the Zygnemaphyceae (after Mix 1972, 1975).

Family	Cell Wall Features
Mesotaeniaceae and Zygnemataceae	1 Cell wall consisting of one piece 2 No shedding of primary wall 3 Outer hyaline (mucous) layer smooth 4 No or only weak incrustations 5 No pores
Gonatozygaceae Peniaceae Closteriaceae	1 Wall may be formed in several segments, that are divided by only very slight constrictions 2 No shedding of primary wall 3 Compact, structured outer layer (warts, spines and ridges originating from outer layer) 4 Differences in nature of ornamentation 5 Pores or pore-like gaps present only in outer layer.
Desmidiaceae	1 Wall consisting of two segments divided by a constriction (isthmus) 2 Primary wall shed 3 No continuous outer layer but mucilaginous envelope originating from pore-organs 4 Stongly ornamented in many cases (originates from secondary wall) 5 Pores which occur in secondary wall

surface or arranged in a regular pattern especially on the cell apex. The
pores contain a mucilagenous plug, the pore apparatus, and Mix (1975)
suggests that this should be regarded as a structure homologous with the
compact hyaline layer of the Mesotaeniaceae on the one hand, or the
Closteriaceae and Gonatozygonaceae (see Table 3) on the other.

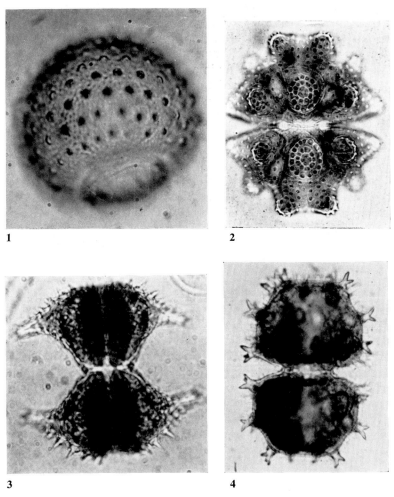

Fig. 38. Warts, granules, simple and bifurcate spines produced by the secondary walls
of Cosmariae. 1 *Cosmarium* sp., an empty semicell (isthmus below) showing its
elaborate surface decoration of large warts and smaller interposed granules.
2 *Euastrum verrucosum* Ehr. showing markedly granulate cell wall with prominent
verrucose protuberances across the broadest part of each semicell above the isthmus
(photomicrograph by Dr Hilda Canter-Lund). 3 *Staurastrum sebaldi* var. *ornatum*
Nordst. showing the cells elaborate decoration of strongly developed apical spines.
4 *Xanthidium armatum* var. *fissum* Nordst. showing the body of each semicell
ornamented with stout, furcate spines.

At the isthmus the structure of the wall of the Cosmariae clearly indicates which of adjacent semicells is the younger since there is always a lip attached to the older one (Pickett-Heaps 1972, his Fig. 16).

Granules, spines or verrucae of varying size and elaboration are characteristic of the cell wall ornamentation of many desmids of the Cosmariae (Fig. 38). Because they are arranged in regular and consistent patterns and can be clearly seen by light microscopy, they are of significance in desmid taxonomy. However, they are not structures produced from the outer wall layer as is the ornamentation of the genera *Closterium*, *Penium* or *Gonatozygon*, but result from the differentation of primary and secondary walls being initiated during the later stages of cell enlargement (Kiermayer 1964, Kies 1968). Thus they are intimately associated with the development of the cell wall proper (Fig. 36: 3).

THE PORE ORGAN AND MUCILAGE PRODUCTION

Pores have never been found in the walls of any of the Mesotaeniaceae, although there is a copious production of mucilage by these desmids.

A detailed study of the structure of the pores so characteristic of the desmid cell wall (Fig. 40: 1–6) was first undertaken by Lütkemüller (1902). He showed in the Cosmariae that in the inner secondary wall the pore is a simple canal in which internally there is some slight enlargement, while externally he described a specially differentiated non-cellulosic cylindrical zone. It was he who termed the whole structure the 'pore organ' and showed that at the inner surface of the pore entrance there are button-shaped swellings. Externally the pore organ quite commonly flares outwards to produce a cone or funnel-shaped structure and this Lütkemüller indicated to be of a complex form. Especially in large desmids numerous smaller pores occur in a regular pattern between the larger complex ones. These are said to be confined to the outer layers of the wall. From the pore opening, pin-shaped or wedge-shaped filaments of mucilage are extruded (Fig. 39). In many desmids there is, in addition, the secretion of a substantial envelope which internally may exhibit a prismatic or fibrillar structure enveloping the cell completely (Figs 40 and 41). As suggested by Fritsch (1935) this no doubt corresponds to the exudation of mucilage through individual pores in the form of a number of closely opposed prisms. Where mucilage masses are recognisable exterior to the pore organs, each occupies the middle of one of the prisms. These prismatic structures have been the subject of detailed examination by both light and electron microscopy by Drawert and Metzner-Küster (1961) and Drawert and Mix (1961) in several desmids, but particularly in *Hyalotheca dissiliens*

(Fig. 40). At the ultrastructural level, Pickett-Heaps (1972, Fig. 45) showed that irregular fibrous material usually remained attached to the external orifice of the pore, and internally there is always a prominent hemispherical invagination of the plasmalemma containing strands of (mucilaginous) material.

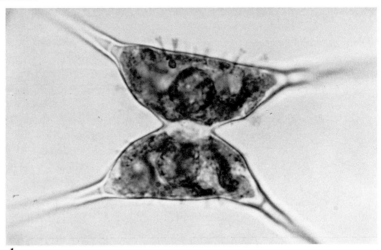

1

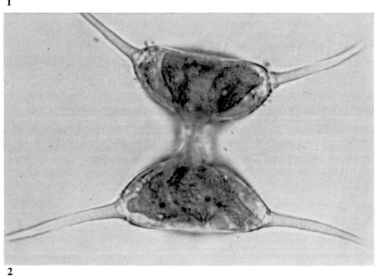

2

Fig. 39. Pin- or wedge-shaped extrusions of mucilage from pores of: 1 *Staurodesmus sellatus* Teil.; 2 *S. mamillifatus* (Nordst.) Teil. In genus *Staurodesmus* the mucilage pores are restricted in their disposition and are a character of taxonomic significance as for example in *S. mamillifatus* (Nordst.) Teil. where they are restricted to a double ring round the apex of the angles of each semicell. (Photomicrographs by Dr Hilda Canter-Lund.)

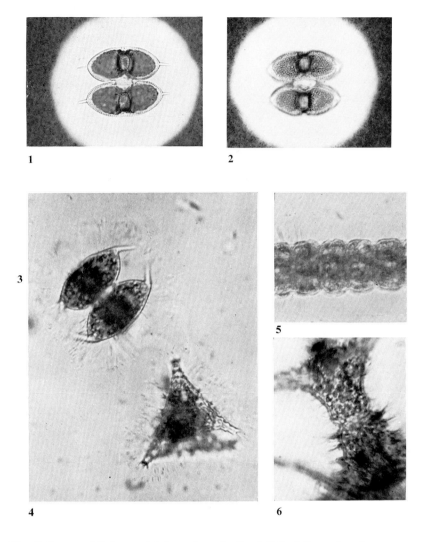

Fig. 40. Pores and their associated mucilage sheaths. 1 Cell of *Cosmarium depressum* (Naeg.) Lund. with contracted chloroplasts in a plane of focus such that pores can be seen traversing the cell wall; 2 surface view of an empty cell of *Staurodesmus dickei* var. *circulare* Turn. showing the cell wall perforated by a well-marked system of pores; 3 *Staurodesmus megacanthus* in side view; and 4 *Staurastrum anatinum* fo. *vestitum* (Ralfs) Brook in apical view from a phytoplankton showing in each a clear enveloping mucilage sheath with a distinctive fibrillar structure; 5 the same fibrillar structure in the mucilage envelope surrounding a filament of *Hyalotheca dissilieus* (J. E. Smith) Bréb.; 6 body of *Staurastrum sebaldi* var. *ornatum* fo. *planctonicum* Teiling showing extruded, wedge-shaped filaments of mucilage stained with methyl violet.

Mucilage extrusions can often be seen without staining, but their prismatic or fibrillar structure becomes much more apparent when stained with methyl violet. They are particularly common and well developed in planktonic desmids and in many filamentous genera such as *Hyalotheca* (Fig. 40). The extent to which mucilage is produced is variable and useful research could be undertaken to explore the environmental conditions which might stimulate, or limit its production.

Although pores were initially considered to occur mainly in the walls of placoderm desmids of the Cosmariae, light microscopy could not show unambiguously whether they were present in all species of *Closterium*, *Penium* or *Gonatozygon*. The EM, however, has demonstrated their undisputed occurrence in *Closterium* and *Gonatozygon* (Mix 1969) while the mesh-like gaps in the outer wall layers of *Penium* spp. are regarded by Mix (1975) as pore equivalents.

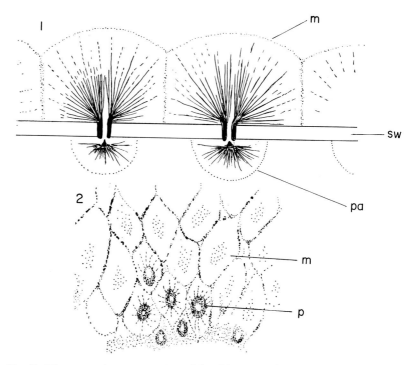

Fig. 41. Diagrammatic representation of desmid pores to illustrate the prismatic nature of the extruded mucilage. 1 L.S. of pores of *Micrasterias rotata* (Grev.) Ralfs showing the pore apparatus or organ and the fibrillar nature of the extruded mucilage. 2 Tangential section of pores thus showing the pores and the mucilage secreted from them at different levels (after Drawert and Metzner-Küster). m = mucilage; p = pore aperture; pa = pore apparatus; sw = secondary cell wall.

These EM studies have revealed two distinct types of pore systems, the *Cosmarium* type and the *Closterium* type. In the former the pore channel extends through the whole thickness of the secondary wall, but in the *Closterium* type it is limited to the outer layer. In this type the underlying fibrillar layers seem to be interrupted and less strongly developed. Thus seen from the perspective of the protoplast beneath, the wall is of a sieve-like nature.

Gonatozygon has pores comparable with those of *Closterium*, whereas the holes in the mesh-like outer layers of *Penium* represent a modification of this pore type. The sieve-like gaps in the outer wall layers of *Closterium*, *Penium* and *Gonatozygon* are thought possibly to have the same function as the continuous pore channels of the Cosmariae in that they are assumed to be points of exchange between the inside and outside of the cell.

In EM study of the filamentous *Bambusina brebissoni*, Gerrath (1975) found that both lateral and apical portions of the cell wall have pores about 100 nm diameter and that the lateral pores travel obliquely through the wall. These are responsible for the production of a 2 µm thick sheath. Gerrath was unable to ascribe any function to the apical pores which, he observed, do not occur at the same places as pores in adjacent cells. (See p. 103 for other details of wall structure in this desmid.)

EM studies by Lott, Harris and Turner (1972) of the walls of *Cosmarium botrytis* have revealed some additional details about the pore system. Thus within the pore bulb they found a number of small fibrils radiating towards the plasmalemma from a net-like fibril layer. These fibrillar layers in turn merge into a markedly electron-dense region at the base of the pore. In oblique sections the same regions of the pore bulb are visible.

The pore sheaths of *Cosmarium* were examined in this same study by freeze-etching and scanning EM. The resulting micrographs show that the secretion from the pores forms a collar round the pore mouth. Observations of freeze-fractured cells also confirm the findings of others (Gerrath 1969, Drawert and Mix 1961) that these secretions forming the pore sheath are composed of needle-like fibrils which radiate outwards from the pore.

As pointed out recently by Chardard (1977) the principal problems still to be resolved about desmid pores relate to their fine structure, differences in the chemical composition of the components, the role of pores in the life of the cell, and their origin and development. In an attempt to throw light on some of these questions he has undertaken light and electron microscope and cytochemical studies of the walls of *Cosmarium lundellii*, *Micrasterias fimbriata* and *Closterium acerosum*.

In *Cosmarium lundellii* the pores which are evenly distributed over the wall at a density of 33–35 per 100 μm^2 consist of a pore cylinder 500–600 nm diameter surrounding a canal 170–350 nm diameter. New information of special interest obtained from EM pictures of 60 000× magnification shows each canal to be surrounded by a cylinder of dense cellulose microfibrils whose deposition is reminiscent of the annular or spiral thickenings of xylem vessels. Extremely clear details of the structure of the internal pore bulb can be seen in Chardard's sections and these complex structures are said to vary with the age and condition of the cell. Initially they appear smooth in outline and he compares them to a skull (calotte), but as they develop and enlarge they take on the appearance of a spongy, perforated mass increasingly depressing the underlying cytoplasm. In the peri-membranous space, fine fibrils of a pectic composition converge towards the bulb so that this region has the appearance of a pin cushion. The role of the pore bulb would seem without doubt to be that of an 'obturator', allowing mucilage to be extruded, but controlling its discharge through the canal.

Where the pore cylinder opens to the exterior one can see in the high magnification EM sections, bundles of filaments entering the pore orifice and forming a sort of collar. The filaments constituting each collar have been shown to be joined laterally to those of adjacent collars by fine lateral filaments, so that in tangential sections (see Chardard's Fig. 3, p. 244) they appear as a sparse, web-like network. The collar is strictly not a part of the pore cylinder, and indeed is not a permanent structure, for it has been demonstrated that when cells are grown in depleted culture media their fibrils are very short so that the collar becomes restricted to the vicinity of the pore opening. In extreme cases it may be altogether absent.

In *Micrasterias fimbriata* because the wall is thinner than that of *C. lundellii* the pore cylinder is shorter, but, by contrast, the collar and pore bulb are better developed. Also the mucilage extruded by these *Micrasterias* cells is composed of much thicker fibrils, forming a dense network which is deposited directly on the wall.

Various staining techniques using ruthenium red, Alcian blue and iron hydroxylamine, produce positive reactions on the pore collar indicating that it is composed of pectic material. Also somewhat drastic extraction techniques in which the cells have been treated with chloral hydrate confirm these findings. By this latter treatment the collars disappear but the pore bulbs persist, and since the latter also resist potash treatment, it is presumed that these structures are composed of cellulose. Chardard emphasises that when mucilage is being secreted, the bulb is surrounded by pectic material, a fact also confirmed by staining techniques. He also stresses that this research on desmids has clearly shown that pectic substances are not amorphous, as was believed for so long, but are fibrillar and composed of more or less long chains of polyuronic acids.

In this 1977 paper, Chardard also reports the discovery of occasional pores with a single external opening but which possess bifurcating canals and two pore bulbs associated with them.

Other cytochemical tests carried out by Chardard (see also Chardard 1974), such as the Thiery reaction (Thiery 1967), confirm not only the polysaccharide nature of the wall but also its heterogeneous nature, and the fact that a substance of unknown chemical composition, but with an affinity for silver, impregnates certain regions, and in particular, the pore cylinder. The Gemori reaction produces a precipitate of lead phosphate on the surface of the pore collar and in the canal; also in the parietal cytoplasm and in numerous vesicles and the saccules of the Golgi apparatus. Such reactions are said to be clear indications of phosphatase activity, not only on the surface of the collar but in the pore canal, and point to the excretion of this enzyme by the cell and presumably its adsorption by its mucilage (see also p. 65).

As for the origin and development of desmid pores, Chardard admits that much still remains to be discovered. However, he reports that they appear in the secondary wall early in its deposition, and in the very brief period before it is shed, when the primary wall overlies the secondary wall, the pores are clearly not functional. Thus they possess no collar and the pore canal is empty (see, however, Pickett-Heaps 1974). As to their inception, Chardard favours the hypothesis that the pores occur at those points on the wall where there is a lack of deposition of wall material, though he suggests as an alternative, that they could be formed following wall erosion by enzyme activity. In either case, the underlying cytoplasm must play a precise and active role, since the position of the pores is so regularly and precisely determined.

In this study Chardard drew attention to the differences in the structure of the pores of *Cosmarium* and *Micrasterias* on the one hand and those of *Closterium* on the other. Thus in *Cl. acerosum* it is especially noticeable that there is no pore collar, and 'mucilage prisms' are never observed. Although Mix (1969, 1972) considers that *Closterium* pores are true pores, Chardard doubts whether this interpretation is valid and outlines the evidence for the existence of 'clear zones', which are sieve-like and traversed by microfibrils but which lack what he terms a 'cement substance' present in other parts of the wall. He suggests that the role of these microfibrils in the clear zones is analogous to that of the pore bulbs of the Cosmariae and are places where molecular exchanges between the interior and exterior of the cell are facilitated. The essential difference between *Closterium* and the Cosmariae is in the inability of the former to extrude mucilage over its entire surface. Only the true pores which occur at the two apices of each *Closterium* cell produce mucilage as filamentous extrusions and these are associated with cell locomotion.

Other observations of interest in Chardard's pore investigation are at the light microscope level and concern the action of Alcian blue at different pH values, which permits the differentiation of acid and sulphated polysaccharides. At pH 0.5 the carboxyl groups do not dissociate as well as the Alcian blue, which complexes exclusively with the sulphate groups, whereas at pH 2.5 Alcian yellow complexes with the carboxyl groups which are formed during the earlier treatment. Thus *Cosmarium* cells treated in this way show a yellowing of the walls with blue dots corresponding to the pores while with *Closterium* one sees a wall coloured uniformly green (the superposition of blue and yellow). This means that the mucilage produced from the pores of *Cosmarium* is composed of sulphated polysaccharides. In these same cells, a lighter blue border on the internal part of the wall is indicative of an accumulation of pectic substances within the perimembrane space.

Chardard points out that much research on higher plants seems now to confirm that the cellulose microfibrils of the wall are synthesised with the guidance of polymerases produced in the plasmalemma and then formed towards the outside of the cell (Vian and Rolland 1972). In lower plants, however, the situation is by no means as clear and there is incontestable evidence of cases where cellulose fibrils (or cellulose proteins) are produced by the Golgi vesicles i.e. within the cytoplasm. He believes that in the desmids there is a double secretion; one exogenous, of mucilage which leaves the cell by way of the pores; the other of a substance involved in the production of the wall. In this respect he would seem to be restating Kiermayer's (1970) observations in his study of cytomorphogenesis in *Micrasterias denticulata*. In this he observed two different types of vesicles pinched off by cisternae of the dictyosomes, large (0.3–1.0 μm) and small (80–250 nm). The former he assumed to contain slime secreted through the pores, while the latter contained wall material.

Lacalli (1974b) has produced a useful review concerning Golgi function and wall formation in the Zygnemaphyceae in which he points out that all the members of the group examined so far contain, as Kiermayer (1970) had shown earlier, two distinct types of cytoplasmic vesicles. These are the large type with diffuse contents and delicate membranes and the relatively small ones with dense or fibrillar contents (see also Staehelin and Kiermayer 1973). The former contain a mucilage which would seem to be important, especially during conjugation, but also in producing the slime layer which surrounds, to a greater or lesser extent, the walls of all vegetative cells. In the instances in which it has been observed that there has been a very considerable vesicle contribution to the growing cell wall it is the small vesicles which have been implicated. As Lacalli states, their contribution in *Micrasterias* cells is to the primary wall, and specifically to the hemicellulosic matrix.

Flattened vesicles with thickened membranes have been observed in EM studies of *M. denticulata* (Dobberstein and Kiermayer 1972, Kiermayer and Dobberstein 1973), *in M. papillifera* (Kies 1970) and in *Cosmarium botrytis* (Pickett-Heaps 1972b). Thus they may well be present in all placoderm desmids with distinct secondary walls. These vesicles are said to arise at the distal ends of dictyosome cisternae and to carry particles on the inner surfaces of their membranes that may take part in the massive deposition of microfibrils which accompanies secondary wall formation. Lacalli (1974b, Fig. 4 and pp. 130–132) illustrates and describes very clearly the possible modes of formation of the large, small and flattened vesicles from the dictyosome cisternae in *Micrasterias*, and outlines how each shows a distinct cycle of activity related to the division activity of the cell (see also Lacalli 1973 and Pickett-Heaps 1972b).

That peroxisomes (microbodies) occur in desmids was first demonstrated in *M. fimbriata* by Gouhier and Tourte (1969) and then by Kiermayer (1970) in *M. denticulata*. Tourte (1971, 1972) has shown them to be present during all stages of the life cycle of the former species. In fully developed, mature semicells they appear to occur most abundantly near the chloroplast and cell wall. Their numbers increase 6–12 hours prior to cell division, and then as daughter semicells are formed these organelles flow ahead of the chloroplast (see p. 117 and Figs 47 : 5 and 51) as it increasingly protrudes from the old into the young, expanding semicell. In the same study, Tourte has demonstrated catalase activity in the peroxisomes indicating that they are involved in photorespiration. She adds, however, that the peroxisomes may also play some role in wall formation or mucilage secretion since they have been shown to play a role in the synthesis of glucosides; also that catalase activity appears to be associated with the development of the Golgi apparatus.

How the cell controls the synthesis and secretion of mucilage remains to be discovered. It is possible, according to Chardard, that microtubules which are known to occur in the cell (see p. 107) participate in the process, perhaps even functioning as a sort of barrier controlling these secretions. Such microtubules are rare, if not absent, in the cytoplasm beneath the pore bulb.

Mucilage extruded from pores may confer on cells the benefits of an envelope of trapped soluble nutrients adjacent to them (Mix 1966). Whether such an enriched environment in fact exists around cells has yet to be proved. Along similar lines, the suggestion has been made that enveloping mucilage may absorb proteins (Yeh and Gibor 1970) since they have shown that bovine serum-albumin in Indian ink becomes deposited on the mucilage trails of *Closterium*. They point out that the absorption of protein may be of ecological significance in that mucilage may act as a site for the concentration of soluble nutrients. However, *Closterium* usually moves away from its extruded trails and so from this supposed potential nutrient source.

CELL WALL TYPES AND DESMID CLASSIFICATION

The EM studies of the desmid cell wall by Mix (1972) have led to a clarification of some aspects of classification (see also p. 85) particularly with respect to the somewhat controversial status of *Gonatozygon*. Until Růžička (1970) suggested this desmid's affinity with the placoderms, (Desmidiaceae) they had, according to several prominent algologists, close affinities with the Mesotaeniaceae and Zygnemataceae (West and West 1904, Bourrelly 1967). However, Růžička's view has clearly been supported by Mix's finding and she has proposed, in consequence, the following desmid classification based on wall features (see also Table 3).

> CLASS: ZYGNEMAPHYCEAE (CONJUGATOPHYCEAE)
> Order I: Zygnematales (Saccodermae)
>   Family 1: Mesotaeniaceae
>   Family 2: Zygnemataceae
> Order II: Desmidiales (Placodermae)
> Suborder I: Archidesmidiinae
>   Family 3: Gonatozygaceae
>   Family 4: Peniaceae
>   Family 5: Closteriaceae
> Suborder II: Desmidiinae
>   Family 6: Desmidiaceae

PORES AND CONNECTING STRANDS

The decisive taxonomic character of the colonial genus *Cosmocladium* is that its cells possess one or more strands connecting adjacent cells of a colony (Fig. 42). These strands are described as arising from a group of pores above the isthmus of each semicell (see p. 16) (Schröder 1900, 1902 and Lütkemüller 1902). The structure of these connecting strands in *C. saxonicum*, the type species, has been examined by Gerrath (1970) by both interference microscopy and EM. By the former, the strands can be seen distinctly as straight rod-like structures 1.5–2 μm wide and up to 50 μm long with a conspicuous enlargement in the middle of each. In the case of cells near the middle of a colony there may be 2, 3 or 4 pairs of strands to each cell. He was also able to show numerous sheath prisms radiating from pores in the apical region. On staining with methyl violet, two groups of pores each containing 12–15 pores either in a broad band or arranged in a circle were visible, though the connecting strands did not stain. The connecting strand was found to be associated with the pore group of one semicell only. Additional sets of connecting strands were found to connect with the same group of pores.

Hiemans (1935) examining *C. pusillum* and *C. subramosum* concluded that the connecting strands were comparable to the connecting strands in the Dictyosphaeriaceae of the Chlorococcales, being derived from the primary semicell wall as it was discarded, following cell division. Gerrath found no evidence of this. He demonstrated that after cell division a new pair of strands is visible joining the daughter cells, and that new strands attach at the same point as the other strands that are present. In consequence, if Hiemans' observations are correct, he questions whether these two species are in fact species of *Cosmocladium*. Clearly they should be reinvestigated.

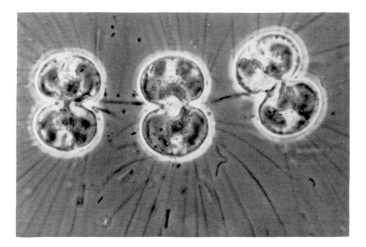

Fig. 42. Phase-contrast photomicrograph of part of a colony of *Cosmocladium saxonicum* de Bary, showing pairs of connecting mucilage strands and numerous sheath prisms radiating from pores in the apical periphery of each semicell (photo by Dr Hilda Canter-Lund).

EM examination of the connecting strands in TS showed them to consist of 12–15 subunits, each of microfibrils arranged in a ring with a less dense central region. Towards the cell, the subunit strands appeared solid. Gerrath has shown that in the isthmal pore group the plasma membrane is withdrawn from the cell wall to form a series of interconnected pore bulbs. Branched or unbranched microfibrils were observed attached to the plasma membrane, which extended perpendicular to the membrane surfaces. For other aspects of the cell wall, and especially its development, see Chapter 5, pp. 110 and 117.

CHAPTER 5
CELL DIVISION

INTRODUCTION

As unique and intriguing as any aspect of desmid biology is the process of cell division, especially its mode of occurrence in placoderms. In the latter it has very special interest in that some of the problems it raises may help in providing a better understanding of some of the principles of morphogenesis and the control of the shape of cells in general. This is because with each division of these often elaborately shaped desmids, two daughter semicells are produced which in the space of a few hours reform the often, extremely complex, symmetrical shape of the parent semicells.

I. DIVISION IN THE SACCODERM DESMIDS

Many members of this family form extensive gelatinous strata in damp terrestrial habitats (e.g. *Cylindrocystis, Mesotaenium* spp.). Their often prolific multiplication takes place following a considerable elongation of the chloroplast and pyrenoid in each semicell. According to Carter (1920) in studies of the process in *Cylindrocystis crassa* and *Netrium oblongum*, the onset of cell division can be recognised by the appearance of a constriction of chloroplast and pyrenoid about half-way between the cell apices. There is then a mitotic division of the nucleus during which time chloroplasts and pyrenoids complete their division. According to Karsten (1918), the period of maximum mitotic activity in *Mesotaenium endlicherianum* is from 11.00 to 13.00 hours, whereas Kaufman (1914) found that in *Cylindrocystis brebissonii* it also occurred most frequently around midnight. Following the division of the nucleus a septum forms across the middle of the cell. This arises as an annular ingrowth from the cell wall which, as it develops, gradually separates the newly formed daughter nuclei which then migrate in opposite directions, eventually coming to rest in the gaps between the, by now, divided chloroplasts. In *Roya* they presumably lie in the notch of the single chloroplast. Carter comments that the chloroplasts apparently complete their division without being closely associated with the nucleus. In this respect the division of saccoderm cells is significantly different from such unconstricted placoderms as *Hyalotheca* and *Closterium* (see pp. 101 and 104).

In some saccoderms the newly formed cells separate soon after division by the dissolution of the middle lamella. However, in those species which tend to occur as groups of individuals held together within a common mucilage envelope, the mucilage enclosing them may show distinct evidence of stratification. This is said to be the result of the gelatinisation of the outer membranes of successive generations. Attention has been drawn by Pumaly (1923) to the fact that in *Cylindrocystis crassa* cell division takes place successively in two or three planes at right angles to one another. This he considers to be a primitive characteristic, and indeed it occurs most commonly in some of the simplest genera of blue-green algae, such as *Chroococcus* and *Gloeocapsa*.

The most detailed account of cell division in the saccoderm desmids relates to its occurrence in *Netrium digitus* and was provided by Biebel (1964). The initial stages are marked by the central region between the chloroplasts, which contains the nucleus, becoming enlarged and 'a short, dim, dark line becomes visible'. Mitosis occurs, and on its completion the new wall, which will eventually divide the cell, forms as an annular ingrowth gradually cutting the daughter cells into two. After their separation the initially flat new wall of each bulges outwards to form the characteristically shaped apex of the cells. The daughter cells then move apart as a consequence, it is suggested, of unequal mucilage production from the adjacent ends of each cell. Chloroplast division in each daughter cell begins some time (1 hour in the case of one strain) after the cells separate and is first noticeable as a slight constriction and is completed within about 10 hours. Details of mitosis in 3 distinct clones of *N. digitus* and in one clone of *N. nagelii* have been reported by Brandham (1964) (see p. 169) (cf. also Maguitt 1925, Kreiger 1933, King 1954).

Few recent detailed studies have been made of cell division of saccoderm desmids and, as noted by Pickett-Heaps (1972), such investigations may produce interesting and significant results.

II. DIVISION IN THE PLACODERM DESMIDS

For many placoderm desmids cell division seems to be the only means of reproduction and many populations, especially in the phytoplankton, are considered to be clones of immense age (Brook 1959a), for in many such species sexual reproduction has never been observed despite detailed observations of them over many years.

There appear to be four categories of cell division amongst the placoderm desmids typified by:

1 *Hyalotheca*
2 *Bambusina*
3 *Closterium*
4 *Cosmarium*

1 HYALOTHECA TYPE

The cells of the filamentous desmid *Hyalotheca* are in all species unconstricted, and as described by Hauptfleisch (1888) division is initiated by the development of a cylindrical strip of membrane on the inner side of the wall at the level of the junction of the two semicells (Fig. 43: 1). From this strip the septum arises as an annular ingrowth and when fully closed across the cell it splits into two, so that each original semicell acquires a new daughter semicell which gradually extends longitudinally until it reaches its mature size (Fig. 43: 2).

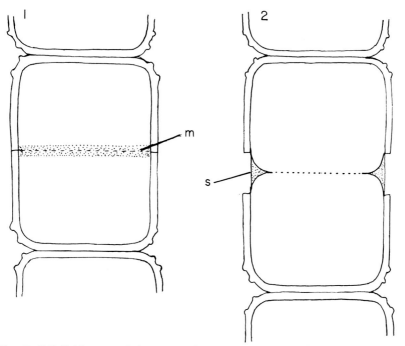

Fig. 43. Cell division—*Hyalotheca* type. Diagrammatic representation of: 1 the cylindrical membrane (m) encircling the inner surface of the cell wall; 2 the development of the septum (s) during cell extension in *Hyalotheca mucosa* (Mert.) Ehr.

According to Acton (1916), when the nucleus divides the resulting daughter nuclei are separated by the transversely encroaching septum. The nuclei then become amoeboid, and each takes up a lateral position opposite the central pyrenoid of one of the chloroplasts. Under the presumed influence of the nucleus, the chloroplast and pyrenoid then divide and subsequently the nucleus moves between the two halves of the chloroplast as these separate. It would be instructive to explore these cytological events at the submicroscopic level in *Hyalotheca*, as has been done for *Closterium* and *Cosmarium* (see below).

2 BAMBUSINA TYPE

In the filamentous desmids *Bambusina*, *Desmidium*, *Haplozyga* and *Streptonema* the cells are separated by what are termed replicate septa in which the adjoining walls develop a cylindrical ring-like ingrowth similar to that found in certain species of *Spirogyra* (Zygnemataceae) (Fig. 44: 1).

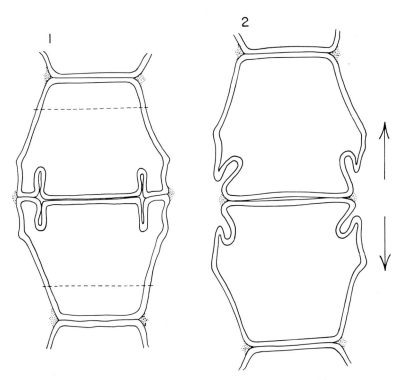

Fig. 44. Cell division—*Bambusina* type. Diagrammatic representation showing: 1 replicate septum; and 2 its unfolding during cell extension in *B. brebissonii* Kütz.

Details of cell division in *Bambusina brebissonii* have been studied by EM by Gerrath (1975) who has provided valuable observations about the ultrastructure of the fully developed division septum, and of the stages in the expansion of the newly forming semicells. He has found that the lateral region of the parent cell is about 200 nm thick, except for a thickened slightly protruding ridge round the cell apex. Adjacent cells are separated by an electron transparent area filled with the remains of the previous initial division septum. Gerrath also records that occasionally a small layer of initial wall from the previous cell division can be seen clinging to the wall, usually near the cell apex.

In common with the basic structure of all other cells of the Cosmariae it was shown that the semicell walls overlap at the isthmus and that microtubules are present here below the plasmalemma.

The division septum consists of 3 distinct parts, of which the centre is the narrow primary wall which is laid down immediately following cytokinesis. Secondary wall material is laid down on both sides of the primary wall. The positions of pores soon become apparent, and each is filled with a granular mass, less electron dense than either primary or secondary walls. Pores appear to be regularly distributed on the folds of the septum and on the exposed lateral portion of the new semicell (see also p. 91).

The original study of cell division in *Bambusina* by Hauptfleisch (1888) indicated that the folded part was the first region to separate as each new semicell began to assume its adult form and that they continued to separate except in the region of the cell apex. Contrary to these observations Gerrath's study indicates that separation appears first at the periphery and then moves to the folds. He has also shown that the primary wall splits into two more or less equal parts with each layer remaining attached to the secondary semicell wall. The pores lose their individual plugs as soon as they become exposed after wall unfolding and then begin to secrete a mucilage sheath.

Gerrath's study is especially interesting in that it shows very clearly the positions of the original bends in the division septum on the mature semicell wall. Thus the apex of the fold is visible as a slight depression towards the semicell apex, the outer margin as a thick ridge near the isthmus, while the remaining angle is also visible as a slight ridge on the lateral wall (Fig. 14: 1).

In discussing the distinctive method of cell division, such as outlined above, involving the formation of a septum with infolded mature walls Gerrath suggests that, possibly because they all form replicate division-septa, *Bambusina*, *Desmidium*, *Haplozyga* and *Streptonema* might be segregated taxonomically and placed into a separate subfamily. He stresses the fact that in these desmids the initial walls do not split until the mature walls have formed, and so differ from the majority of placoderms where splitting occurs soon after formation and mature walls are not laid down until new semicell expansion is almost complete (cf. p. 91). In the *Bambusina* type of division it is quite clear that the wall must remain flexible until unfolding occurs. Gerrath speculates as to why *Bambusina* 'et alia' have evolved what he terms an 'essentially internal prefabrication of the new semicell'. He questions the adaptive advantage conferred by this so-called experiment in cell morphogenesis, and wonders if it might possibly minimize the vulnerability of the cell in that it does not have to exist, as do other placoderm desmids, with only a flimsy primary wall until full cell expansion is achieved before the much firmer secondary wall is laid down.

3 CLOSTERIUM TYPE

Several investigators have described cell division in *Closterium* (Lütke-müller 1902, Lutman 1911, van Wisselingh 1913, Brandham 1964) and the process has been concisely summarized by Fritsch (1935), including some discussion of discrepancies in the interpretation of events (Figs 45 and 46). More recently Pickett-Heaps and Fowke (1970) have studied the process at the ultrastructure level in *Closterium littorale*, with special reference to the cytological changes associated with mitosis.

The period of maximum mitotic activity has been determined for several *Closterium* species as shown in Table 4 (after Brandham 1964).

Table 4. Period of maximum mitotic activity in *Closterium* species.

Author	Species	Times when mitosis found
Fischer 1883	*C. delpontii*	Midnight (cleavage)
Lutman 1912	*C. ehrenbergii*	21.00–22.00 hours
	C. moniliferum	21.00–22.00 hours
Karsten 1918	*C. moniliferum*	Peak 11.00–13.00 hours
Brandham 1964	*C. siliqua*	Peak 08.30 and 22.00 hours

Observations show that as mitosis proceeds the central gap between the chloroplasts of adjacent semicells widens as the spindle elongates (Fig. 46: 2). The annular ingrowing septum, which Fritsch (1935) emphasizes is formed not at the boundary between two semicells, but a little way beyond in the younger semicell, appears in early prophase as a slightly thickened ring. It is suggested that, at this location, the outer wall is lacking and it is only the inner which is thickened to project slightly into the cell cavity. This can be seen by light microscopy as a faint transverse line, for it marks a slight internal constriction of the wall, clearly seen in one of

First division

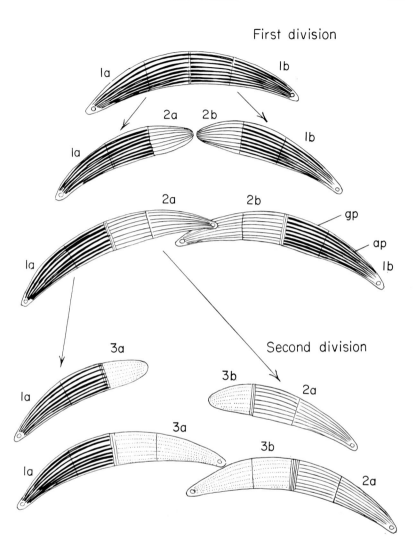

Fig. 45. Cell division—*Closterium* type. Diagram showing two successive divisions of a *Closterium* cell to illustrate the distribution of the parent cell walls to the daughter cells. The walls of the 1st generation cell (1a and 1b) are drawn with heavy longitudinal striations; those of the 2nd generation (2a and 2b) with lighter striations; while those of the 3rd generation (3a and 3b) have striations shown as dotted lines. gp = girdle portion; ap = apical portion.

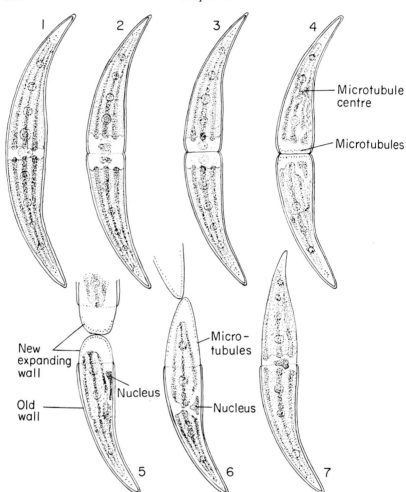

Fig. 46. Semidiagrammatic representation of the cytological events accompanying cell division in *Closterium littorale* Gay. (after Pickett-Heaps and Fowke).

Pickett-Heaps and Fowke's (1970 p. 195, Fig. 14) excellent micrographs taken at the metaphase of mitosis. When division is about to occur this thickened ring becomes drawn out to form a cylindrical strip due to a slight elongation of the cell (Fig. 46: 2). As telophase proceeds the septum grows inwards from this new wall segment, somewhat like the closing of an iris diaphragm. In some species of *Closterium* and also *Penium* (e.g. *C. striolatum* and *P. spirostriolatum*) two or three girdle bands can be seen by successive stretchings of the walls on each side of the line of thickening (Fig. 45: 3b).

In a time-lapse film, by Brandham (1965), of movement in *Closterium acerosum* it is interesting to observe that although movement ceases during cell division it is re-established long before the new semicell has become fully extended, so that the two newly formed cells move apart quite vigorously.

After this cell extension has been achieved, as stressed by Pickett-Heaps and Fowke, the cell 'faces considerable logistical problems in cell reorganisation', preparations for which begin during the late stages in mitosis. What has to be accomplished is the development of a new semicell from each of the mature parent semicells which have been bisected by the growth of the septum. During this morphogenesis two closely coordinated and integrated events occur: first the expansion and development of a cell wall and second, nuclear migration which, as will be explained, is intimately associated with chloroplast cleavage (see Fig. 46).

Pickett-Heaps and Fowke record that after septum formation the daughter nuclei come together on either side of the septum, move to the concave side of the cell and become somewhat distorted along with the chloroplast which they appear to compress. Detailed studies of the cytological events associated with cell division by Brandham (1964) (see p. 162) in *Closterium siliqua* and *C. ehrenbergii* show, contrary to the observations of Pickett-Heaps and Fowke on *C. littorale*, that the nuclei migrate along the concave margin of the cell wall to reach their new positions between the dividing chloroplasts. Pickett-Heaps and Fowke attribute the actual compression, however, to the development of a 'microtubule centre' which migrates towards the pole of each cell until it lodges in the future point of cleavage of the chloroplast midway along its length. After the microtubule centre reaches the point of chloroplast cleavage, the nucleus begins to migrate along the cell, and in the later stages of its migration, the nuclear envelope is pulled into the deepening furrow. Eventually the chloroplast becomes divided into two equal parts and the somewhat compressed nucleus expands between them to achieve its full interphase size (Fig. 46: 5–7).

Although no microtubules are initially associated with the formation of the septum after the daughter nuclei reorganize following mitosis, they approach the thickening septum, and as they do so increasing numbers of transversely orientated microtubules appear adjacent to it. It is suggested that some of these microtubules associated with the septum originate from the spindle.

After the completion of telophase, cell expansion begins to cause the material of the thickening septum to split into two, and eventually a furrow develops and deepens between the two developing semicells. Finally, the two newly formed complete cells separate though they may remain loosely

held together for a time held by a sheath of mucilage. As cell wall development proceeds, cell symmetry is restored and the nucleus takes up its
position in the centre of the cell.

During cell wall expansion the above mentioned septum microtubules
adopt a particularly significant disposition near the new cell wall. They
become arranged in a hoop-like configuration only along the region of the
new wall and this is maintained throughout its expansion. After this is
completed, however, they gradually disappear.

Pickett-Heaps and Fowke suggest that nuclear migration during cell
division, though a unique and hence remarkable phenomenon not found
in any other organism, is probably typical of many unconstricted desmids.
It has also been described in *Pleurotaenium trabecula* by Brandham
(1964). They cite in this connection the observations of Acton (1916) and
Carter (1920) on *Hyalotheca*, *Netrium* and *Cylindrocystis* in which they
suspect there is a microtubule centre similar to that which they have found
in *Closterium*. Their suggestion has yet to be confirmed by ultrastructure
observations of these desmids.

4 COSMARIUM TYPE

This fourth mode of cell division is common to all constricted placoderm
desmids and so occurs not only in all species of *Cosmarium* (Fig. 47) but
also in species ascribed to the genera *Staurastrum* (Fig. 48), *Euastrum*,
Micrasterias etc. The most interesting aspect of the *Cosmarium* type of
division is that, as in *Closterium*, each time a cell divides its symmetry is
completely destroyed by a wall that grows across the junction of adjoining
semicells, which, in such cells, is most commonly a distinctly narrow isthmus (Fig. 47). To restore the cell's symmetry a new semicell has to be
regenerated which, except in cases where there is asymmetry (Chapter 2),
becomes the mirror image of the parent semicell. In addition, of course, all
the major cell organelles have to be reformed, in the correct position,
within the semicell.

As in the case of *Closterium*, Pickett-Heaps (1972) has described cell
division in the widely distributed desmid *Cosmarium botrytis* as it appears
by light microscopy (see also Brandham 1964), and also at the ultrastructural level by EM. His photomicrographs show that even at the very
early stages of division adjacent semicells can be clearly seen to push apart
and presumably this separation is a consequence of the initiation of the
septum. It is of interest to note that Brandham (1964) who has studied in
detail the cytological events associated with cell division in a number of
the Cosmariae reports that in *C. botrytis* the isthmus usually elongates at
metaphase, but sometimes not till anaphase or telophase. When the isth-

mus fails to elongate a peculiar type of binucleate cell results (see also p. 122). In *C. difficile* he found that isthmus elongation occurred at late prophase, as was also the case in *Xanthidium antilopaeum* (Fig. 49) and *Spondylosium secedens*. However, in *S. papillosum* it did not happen until late anaphase. His most interesting observations in respect to isthmus elongation relate to the unusual *Cosmarium pseudoconnatum*, in which

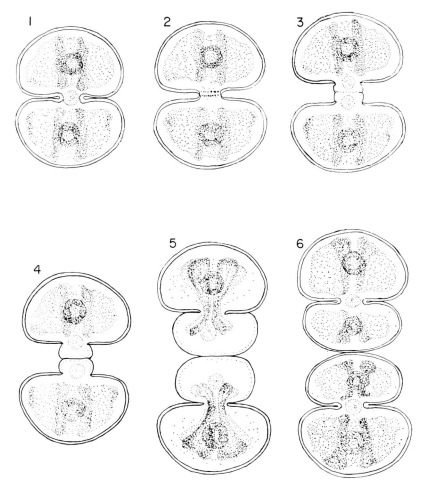

Fig. 47. Semidiagrammatic representation of the major stages in the *Cosmarium* type of cell division. 1 Front view of cell before division; 2 metaphase of mitosis at which time semicells have separated; 3 semicells have pushed further apart and daughter nuclei reforming. Septum almost complete; 4 beginning of expansion of primary wall by which time nuclei are reconstituted; 5 chloroplasts of old semicells start to move into new semicells into which nuclei have also moved; 6 division almost complete and secondary wall deposition under way.

species an isthmus is barely distinguishable. In this desmid the newly formed septum appears to be complete before any elongation of the isthmus occurs. This was also the case with *Pleurotaenium trabecula*.

It has been shown by Kiermayer (1964) that in *Micrasterias denticulata* the septum is of special significance for the determination of cell morphology. Indeed he has provided proof that the septum carries, as a sort of template, the potential pattern for the cells subsequent differentiation (das 'Septum Initaalmuster') (see p. 187). In a later investigation Kiermayer (1967) has treated dividing cells during anaphase to early telophase of mitosis with 10% ethanol for varying period. Septum development, which takes from 15–20 minutes, is associated with active cytoplasmic flow and the movement of numerous granules into the isthmal zone, and when complete appears as a continuous dark line. Immediately after ethanol treatment the plasma flow ceases and during the period of narcosis any further septum development appears to be completely arrested. However, when the ethanol is flushed out with fresh nutrient solution, plasma flow recommences and the structures associated with septum formation may again become recognizeable.

Subsequent septum development, however, depends on the length of the period of narcosis and the stage of development of the septum before ethanol treatment. Thus when a cell with a well developed septum, not quite completely closed, was narcotised for only 10 minutes, closure was

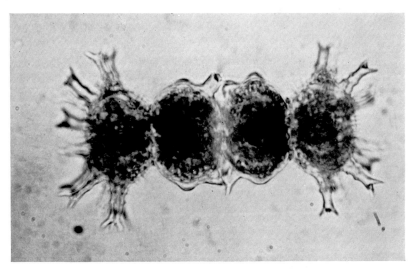

Fig. 48. Cell division in *Staurastrum arctiscon* (Ehr.) Lund. showing the regeneration of daughter semicells and the restoration of cell symmetry. The parent semicells are at the left and right of the picture, the developing daughter semicells in the middle.

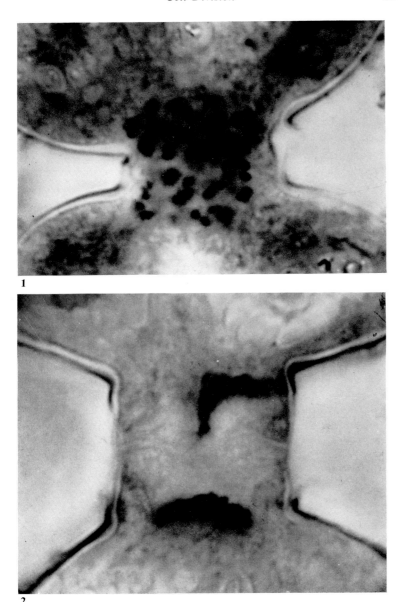

Fig. 49. Mitosis and cytokinesis in *Xanthidium antilopaeum* (Bréb.) Kütz. 1 Late prophase at which stage the isthmus has started to elongate. 2 Late anaphase (the lower plate has been distorted) the continuity of the elongating isthmus with the internal layer of the cell wall can be clearly seen. It will be noted that the septum is virtually complete. (Photomicrographs by Dr P. E. Brandham.)

completed successfully once the treated cells were transferred to nutrient solution. They then proceeded to form complete and normal semicells. However, when cells, whose septum at the time of a 10–min ethanol treatment were still only primarily developed, were transferred to fresh nutrient they did not close completely before the onset of all extension and produced double or giant cells, with a shared middle piece of the type described below (p. 121 and Figs 52 and 53).

Complete and irreversible stoppage of septum development occurred when the period of ethanol narcosis was extended. Thus after 60 minutes of treatment followed by resuspension in nutrient medium the septum showed no further growth. Subsequent development of such cells with open septa also produced double cells. Details of the course of these events are presented in tabular form in Kiermayer's paper (1964). The significance of these experiments on septum formation and cell differentiation will be discussed in Chapter 8 (p. 185) on morphogenesis.

As Lacalli (1973) states, septum formation is still something of a mystery. Centrifugation experiments by Kallio (1951) on *Micrasterias* cells in early stages of mitosis have shown that the septum always forms inside the isthmus quite independently of the position of the mitotic spindle apparatus, and he concludes that 'the prerequisite necessary for the formation of the septum must exist in the ectoplasm or cortical layer adjoining or embedded in the wall of the isthmus'. It was also shown by Kallio (1963) in double cells of *Micrasterias* which have 2 isthmuses, but only one nucleus (see pp. 185 and 187), that the position of the septum does not depend entirely on nuclear position, since a septum can form at both of these isthmuses. Kiermayer (1968) has shown that there is a band of wall microtubules surrounding the isthmus and suggests that this might be responsible for septum initiation. In double cells each isthmus presumably initiates its own septum. In Lacalli's ultrastructural study of cytokinesis in *M. rotata* with its striking EM photomicrographs, he has developed a most instructive semidiagramatic summary of the events leading to septum formation. These figures which are reproduced here (Fig. 50), in somewhat modified form show the following:

1   The interphase condition with the nucleus situated in the isthmus, mid-way between the two semicells, and with wall microtubules apparent, closely adjacent to the cell wall in the cleft formed by the isthmus. In this region, the overlapping walls of adjoining semicells can be seen.

2   The primary wall begins to be deposited during the earliest stages of prophase of nuclear division where it appears as a thin girdle of material. It never appears to extend further from the isthmus than the first pore (see p. 93) and may terminate somewhat abruptly there (see also Fig. 50).

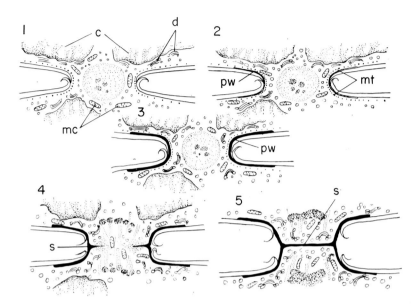

Fig. 50. Cell division in *Micrasterias rotata* (Grev.) Ralfs: a semidiagrammatic representation of the events associated with septum formation shown at the ultra-structural level. 1 Interphase condition; 2 earliest prophase with initial appearance of primary wall as a girdle surrounding the isthmus; 3 later prophase with thickened girdle; 4 early telophase with spindle microtubules and regions of dense cytoplasm at the septum edge; 5 late telophase with the septum complete. (After Lacalli.) c = chloroplast; d = dictyosomes; mt = microtubules; mc = mitochondria; n = nucleus; pw = primary wall; s = septum.

3   The girdle of primary wall material thickens during prophase during which time the isthmus begins to extend. As this occurs, the wall micro-tubules become less numerous.

4   By early telophase, in which spindle microtubules can be seen, the original semicells have separated, though joined by the expanding primary wall from which the septum develops centripetally, gradually cutting the semicells in two. Regions of dense cytoplasm are apparent at the septum edge.

5   By late telophase, the septum of primary wall material has completely divided the two semicells.

By metaphase when the nuclear envelope has dispersed, microtubules become interspersed amongst the chromosomes and typical spindle microtubules extend from small double chromosomes to the poles, or they pass through the metaphase plate. Subsequent mitotic events leading to telophase are much as in higher plants and also the open spindles observed are presumed by Pickett-Heaps (1973b) to be analogous to those higher plants. At telophase the reforming nuclei are initially small, flat and dense but they enlarge as the chromatin disperses. He reports that the way in which the septum forms is unclear, but comments that the aggregated nature of its fibrous material suggests that the latter was added as the contents of discrete vesicles, always numerous and possibly derived from Golgi bodies.

When the completed septum has thickened, the two incipient daughter cells begin their expansion, and as they balloon outwards they inevitably push apart. They remain joined, however, at the septum centre by this shared, but gradually decreasing, region of the wall. The primary wall material becomes thinner during the ballooning process and it is assumed must be added to, to provide the material necessary for its full and considerable expansion. Towards the conclusion of this enlargement Pickett-Heaps' micrographs suggest that the wall material becomes increasingly orientated by its stretching.

Within an expanding semicell the cytoplasm streams vigorously. According to Kiermayer and Jarosch (1962) different directions of such streaming are apparent in growing *Micrasterias rotata* semicells and can be related to particular stages of this desmid's elaborate development. Experimental reduction of cell turgor immediately following septum formation has been shown to lead to a significant interruption of cell extension, but with a corresponding and abnormal increase in cell wall thickness (Kiermayer 1964, 1970). The importance of turgor pressure in cell extension would seem to be indicated by Kallio's observation (1957) that teratogenic double cells (p. 121) occur by incomplete septum closure following cold shock and are caused by abnormally high turgor pressure (see also Kiermayer 1966).

Also pertinent to these observations is the short paper by Kobayashi (1973) on the relationship between cell growth and turgor pressure in *Micrasterias americana*. She calculated the osmotic pressure of cells in terms of their critical plasmolysis at various stages in the growth and expansion of new semicells. Values were determined from the volume ratio of shrunken protoplasts to whole cells in 0.3M mannitol which was considered to be physiologically inactive. The OP in a non-dividing cell was equivalent to 0.2M mannitol but decreased as cell development progressed, reaching 0.11M mannitol when complete. When volume ratios of shrunken protoplasts to old half cells were measured at various stages, a constant value was obtained, indicating that no net change in the amounts of osmotically active substances present in the cell took place during development. Kobayashi argues that as growth proceeds such substances in the cell would be diluted by water, the amount of which is equivalent to the volume of the newly developed semicell. After the completion of development, the OP in the newly divided cells would increase until the next cell division. As she states, 'the question of when or how this occurs remains unresolved because it is very difficult to determine the ages of such cells'. (See also Ueda and Yoshioka 1976.)

Growth rates of *M. americana* in various concentrations of mannitol at 20°C were found to differ insignificantly in solutions containing 0–0.08M. However, it decreased considerably in 0.12M and was almost zero in 0.16M mannitol. As Kobayashi points out many reports emphasise the sensitive nature of many plant cells to water deficiency and so it is remarkable that growth rates of *Micrasterias* are not noticeably retarded even at such high osmotic concentrations as 0.08M mannitol. She also stresses that contrary to Kiermayer's observations (1962, 1964) cell wall thickening of new semi-cells was not apparent in 0.08 or 0.12M mannitol solutions, though in 0.16 or 0.20M mannitol uneven wall thickening was observed. At 0.25M concentration there was no wall thickening, though 'layers of deposits were found outside the plasma membrane which had been separated from the cell wall due to plasmolysis'. These observations point to the conclusion that cell wall synthesis can proceed even during plasmolysis.

When almost fully expanded, regular indentations begin to appear in the new semicells along their outside surfaces and these are the precursors of the characteristic cell ornamentation that develops as the secondary wall matures. Pickett-Heaps (1973b) suggests that these protrusions could have arisen as a consequence of alternate thick and thin regions of the wall, the orientation of the microfibrils being more random in the thicker than the thinner, where they have a distinct orientation. Also, he suggests that it may be significant that the plasmalemma usually remains tightly adpressed to the thicker regions while it lies at some distance from the adjacent thin portions of wall.

The daughter nuclei always remain in the new expanding semicells, and when the latter are approximately half formed the chloroplast begins to move through the isthmus of each parent semicell (Carter 1919, 1920), though the question remains as to how it moves. However, it has been demonstrated that both colchicine and trifluoraline, chemicals which disrupt microtubules, prevent the migration of the chloroplast during cell division in *Micrasterias* (Lehtonen 1977, see also p. 191).

As stated previously (p. 108), the nuclear migration associated with cell division and the reorganization of the new semicell is an event unique to desmids. In Brandham's study of division in a range of Cosmariae he found *Staurodesmus convergens* to be an especially favourable species for the observation of this phenomenon; and also of chloroplast division during the growth of the new semicell. Thus, he found that at an early stage each daughter nucleus is carried into the new semicell by the enlarging

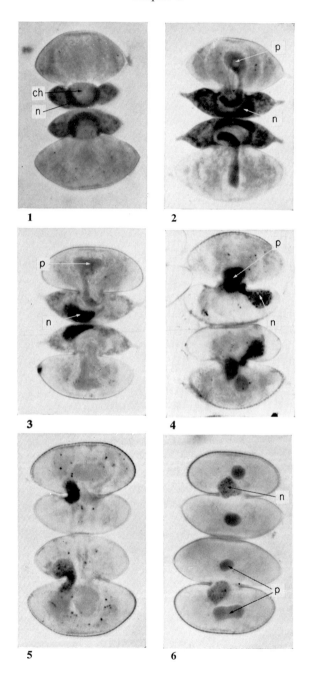

chloroplast (Fig. 51: 1), which once it passes through the isthmus its lobes curve outwards to touch and then slide along the expanding cell wall (Fig. 47: 5). As this event proceeds, the pyrenoid enlarges through the isthmus and eventually divides (Fig. 57: 2–6), although the division may in some cases be delayed. As the young semicells become increasingly filled with chloroplast material, the daughter nuclei begin to migrate around the lateral margin of the chloroplast towards the isthmus (Fig. 51: 3–5). Brandham comments that this, however, differs from most of the Cosmariae in which migration is between the chloroplast and the front wall of the semicell. When the nucleus arrives at the isthmus, the chloroplast divides and it then can assume its normal position in the semicell (Fig. 51: 6). No distinctive cytoplasmic structures have been found which appear to be associated with either chloroplast cleavage or nuclear migration.

When the primary wall is almost fully expanded the secondary wall begins to be laid down with its pattern of ornamentation matching that already established in the primary wall. The secondary wall also acquires its system of mucilage pores, the sites of which are indicated very early in wall deposition by elongated cylindrical plugs of homogeneous material which penetrate the entire secondary wall. On the inner side of the wall the end of each plug is dome-shaped and surmounted by a region of tightly applied plasmalemma. The wall also thickens and becomes cone-shaped around the internal orifice of each future pore. Pickett-Heaps suggests that for the pores to become functional the plug material has somehow to be eroded, and he has observed in this connection that as the secondary wall thickens the plug material becomes paler and increasingly distinctive in its appearance as compared to the wall.

Daughter cells remain joined to one another, apex to apex, until the casting off or 'shedding' of the primary wall, which can often be seen in cultures of desmids near recently divided cells (Fig. 35). Brandham's (1965) time-lapse films show that the daughter cells may then move apart quite vigorously, presumably due to the extrusion of mucilage.

Fig. 51. Nuclear migration during cell division of *Staurodesmus convergens* (Ehr.) Florin. 1 Early stage in enlargement of daughter semicells. The nuclei are carried into the new semicells by the enlarging chloroplasts. 2 and 3 Later stages in semicell growth showing the pyrenoids elongating into the daughter semicells prior to their division. In 3 the nuclei have been pushed to one side prior to their return to their position within the isthmus. 4 The nuclei migrating round the lateral margin of each daughter semicell. 5 The nuclei have reached the isthmus of each cell but because the chloroplast is still undivided they do not occupy the centre of the cell. 6 The chloroplasts have divided and so the nuclei move to the centre of the isthmus. ch = chloroplast; n = nucleus; p = pyrenoid. (Photomicrographs by Dr P. E. Brandham.)

THE CONTROL OF CELL DIVISION

Fully synchronous divisions of a desmid in culture have been reported for the first time by Schulle (1975). This was achieved with the commonly occurring planktonic desmid *Staurastrum pingue*, and it was found that synchrony could be maintained for as long as required. It was found that cell division always occured at a definite time after the beginning of the light period when light and dark periods were alternated regularly. Such a finding suggests that the onset of illumination triggers the events which set cell division in motion.

Schulle isolated *S. pingue* from an actively growing population in the Titisee and raised cultures in 1% CO_2-enriched modified Rodhe No. 8 solution (Rodhe 1948) with added EDTA. His results show that at both a 16:8-h light:dark and a 12:12-h light:dark regime divisions began to occur consistently some 3–3.5 hours after the beginning of the light period. Thus, he was able to conclude that illumination was the 'time-keeper' for synchronous cell development. His analysis of the observed stages in cell division and subsequent development shows that the initiation of division, as indicated by the formation of a septum at the cell isthmus, reaches a maximum in both light regimes after 6 hours and then declines to zero after 15–16 hours. After this time all divisions ceased, though his graphs suggest that a few divisions were still being initiated in the dark in the 12:12-h light:dark cultures. If one extrapolates to the 17th hour after illumination, by then all stages of development would appear to have been completed. It should be noted that in none of his experiments were all cells in some stage of division or development so that the number of non-dividing cells was never zero.

Schulle reports that the total period of development of newly formed semicells was from 2–3 hours. However, although their development was complete after this time, the actual separation of newly formed cells usually took another 3 hours to achieve. Also, although attempts were made to reduce the generation time for *S. pingue*, it was found that a given cell would, at least under the imposed conditions, divide at best only once every 24 hours.

According to Pirson (1961), who introduced the concept of a synchronization constant (Synchronizationschärfe)

$$K_s = 1 - \frac{t_s}{t_g}$$

in which t is the time for the initiation of synchrony (in the case of *S. pingue* 3–3.5 hours when septum formation was observed) and t_g the total development period (8 hours for *S. pingue*). The K_s for this desmid is of the order of 0.65. This may be compared with 0.88 for *Chlorella fusca* (Soeder, Schultze and Thiele, 1967) and the very high value of 0.96 for *Chlamydomonas moewusii* (Bernstein, 1960).

It should be noted that Biebel (1964) in his study of the sexual cycle of the saccoderm desmid *Netrium digitus* (see pp. 100 and 129) obtained synchronized divisions of many cells when 3–6 week old cultures were transferred to fresh medium and placed in 'bright light'. Divisions began during the 11th hour of illumination on the day following transfer and continued for about a week, occurring at the same time each day.

POLYMORPHISM

The feature which more than any other has attracted algologists to study desmids, especially placoderms, is their considerable morphological variability. As pointed out by Rosenberg (1944) there are two principal reasons for this: first, the high degree of cellular differention which provides many readily recognisable morphological details; second, the special mode of cell division. With reference to the latter, it should now be clear that their unique mode of division theoretically preserves any morphological change which may happen in a developing semicell for an unlimited period. Assuming the survival of all the daughter cells of a population, the mode of division should, moreover, theoretically result in the retention and propagation of every variation that may arise.

In some populations there may be no marked differences in the appearance of the semicells, but very often it is possible to find quite a range of variation to the extreme condition where there is marked asymmetry between adjoining semicells (see pp. 33–38). As long ago as 1915 Ducellier provided a well documented summary of such aberrant forms in different desmid genera and developed the theory that variations may occur when the new semicell is affected by a change in conditions during its formation, while the old semicell retains its original morphology. Another interesting early study of polymorphism and its implications especially for desmid taxonomy is that of Playfair (1910).

More recent detailed studies of naturally occurring populations of various species (or species-groups) of placoderm desmids have been undertaken by various investigators. For example, *Euastrum spinifera* var. *duplo-minor* (Vidyavati and Nizam 1972) *Micrasterias mahabuleshwarensis* f. *wallichii* (Teiling 1956) *Micrasterias* spp. (Tyler 1970), *Xanthidium subhastiferum* (Rosenberg 1944), *Arthrodesmus mucronulatus* (Bicudo 1975), *Cosmarium botrytis* (Kirk and Cox 1975), *Staurastrum furcatum* and *S. inflexum* (Villeret 1951a), *S. sebaldi* var. *ornatum* (Lind and Croasdale 1966), *S. anatinum*, *S. gracile*, *S. cingulum*, *S. pingue*, *S. planctonicum* and

S. chaetoceras (Brook 1959) and *S. freemanii* (Brook and Hine 1966). Such investigations have led to quite substantial revisions in the taxonomy of the desmids involved, especially with reference to many of the so-called varieties ascribed to them. Such studies also point to the fact that much of the currently confused state of desmid taxonomy can only be resolved by similar detailed studies of populations.

Experimental studies of desmid polymorphism in which the cells are grown in controlled culture conditions were first carried out by Ondracek (1936). Unfortunately, he chose to investigate a species which showed little variability in its morphology. Lefèvre (1939) provided an interesting account of desmid variability in culture but probably the most important early investigations was that of Rosenberg (1944) on *Xanthidium subhastiferum*. In this she looked at the development of its spine-like processes which, when fully developed, number 8 (4 pairs) regularly disposed round the margin of each semicell. Many semicells possess less than 8 (often only 4) and there are those in which there is no production of spines so that the cells look like a species of *Cosmarium*. Rosenberg complemented her culture study with observations of the desmid as it occurs in the phytoplankton in the N. basin of Windermere. In agreement with Ondracek's (1936) conclusions Rosenberg states that morphologically homogeneous populations of desmids can be maintained only if their division rate is high, and that as it declines so the proportion of aberrant cells within it increases (Figs 8: 3 and 19: 3, 4).

Reynolds' (1940) paper on seasonal variation in a planktonic population of *Staurastrum* (*paradoxum*) *chaetoceras* has already been alluded to (p. 33). In this study the biradiate form, possessing two long slender processes produced by each semicell, was found to be most frequent in autumn, winter and spring, while the triradiate form predominated in mid-summer. 2 + 3 Janus types (see p. 34) were as to be expected, most common during the two change-over periods, that is spring to summer and summer through to winter.

Apposite to this study is that of Brandham and Godward (1965) on

Table 5. Symmetry of 50 pairs of sister cells from clonal cultures of *S. polymorphum* at 4 temperatures.

Temp (°C)	Symmetry of parent cell with daughter semicells					
	A 3/3–3/3	B 4/3–3/3	C 4/3–3/4	D 3/4–4/3	E 3/4–4/4	F 4/4–4/4
20	40	7	1	—	—	2
15	36	5	—	—	—	9
10	33	—	—	5	—	12
5	27	—	—	2	1	20

the radial symmetry of *Staurastrum polymorphum*, a desmid which naturally occurs in tri- quadri- and even pentaradiate form. The effect of temperature on growth rate and the production of tri- and quadri-radiate 'facies' were followed. They found that although the multiplication of 3/3 and 4/4 type cells was almost equal at 5°C, at higher temperature that of 3/3 cells exceeded the 4/4 cells. As indicated by their results, repro-duced in Table 5, there are 6 possible combinations which can occur in a pair of daughter cells.

The low multiplication rate of 4/4 cells at 20°C is indicated by the prod-uction of types B and C at high temperatures, for these represent an actual change from 4/4 cells to 3/4 and then to 3/3 cells. As Brandham and Godward point out, this phenomenon accounts for the faster rate of 3/3 cells which they observed at higher temperatures. The reverse phenom-enon clearly occurs at low temperatures, where, as can be seen from the table, there is a development of 4/4 cells from 3/3's by way of 3/3 cells. Clearly these observations are in accordance with the field observations of Reynolds (1940) on changes in radiation (see also Brandham 1965 and p. 33). It is of interest to note that in collections of phytoplankton from the Sasumua Reservoir situated at 2480 m in the Abadare Mountains of Kenya, one of the commonest desmids was *Staurastrum sebaldi* var. *ornatum* (Lind and Croasdale 1966). Forms varying in their radiation from biradiate to pentaradiate were found, but the 4/4 facies of this plank-ter was by far the most abundant between November and February usually comprising from 60 to 84% of the population. Presumably water tempera-tures of the reservoir were at their lowest during this period. Unfortunately data were not given about the population structure for the remainder of the year, though the authors of the paper state that, although present through-out, the numbers at other times are always very small.

That the determination of the degree of radial symmetry is dependent on the rate of semicell development gains support from the observations of Starr (1958) and Kallio (1953), who observed, in *Cosmarium turpinii* and *Micrasterias thomasiana* respectively, that diploid cells of these species (see p. 178) commonly show a higher degree of radiation than haploid cells, for diploid cells commonly multiply less rapidly than hap-loids (Kallio 1951). The possible underlying mechanisms involving changes in radial symmetry will be discussed in the chapters on nuclear cytology and morphogenesis (Chapters 7 and 8).

ABERRANT DIVISIONS

In *Cosmarium impressulum* enlarged cells have been observed which contain two nuclei which lie side by side in the isthmus. They arise as a consequence of the nucleus of a normal cell being displaced from the isthmus so that, following cytokinesis, one of the newly formed cells is binucleate, the other enucleate (Kallio 1961). The binucleate cell is effectively diploid, and the new semicell which forms from its subsequent division is significantly larger than normal. This may then give rise to one daughter desmid in which both semicells are larger than normal, and a second daughter cell which is asymmetrical with respect to semicell size (Brandham 1965).

It has also been observed that some semicells fail to separate during mitosis with the result that what should be new semicells do not develop because they cannot enlarge. Although a septum is laid down across the isthmus and becomes very thick, and subsequent mitotic divisions, synchronous in both nuclei, have been seen in them, their further development has not been followed.

It is not uncommon in old cultures of placoderm desmids to find a square or rectangular cell-like body intercalated between two semicells (Fig. 52). These arise because of a partial or complete failure of septum formation following mitosis. The elongated isthmus enlarges but cannot develop into two separate semicells because of the incomplete formation of the septum (Rosenberg 1940). The resulting shape seems to relate directly to the degree of failure of septum formation (Fig. 53). Repeated failure

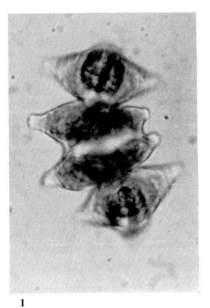 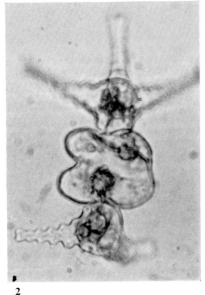

1   2

Fig. 52. Aberrant cell divisions in: 1 *Staurastrum gracile* Ralfs; and 2 *S. pingue* Teiling which have resulted in the production of giant, binucleate cells.

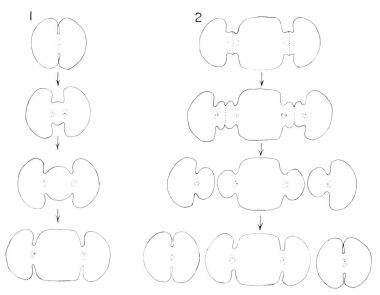

Fig. 53. Diagram illustration: 1 the formation; and 2 the division of giant binucleate cells in the Cosmariae (after Rosenberg). Note that a giant cell produced two normal cells following its division but that the giant cell remains as such in the population.

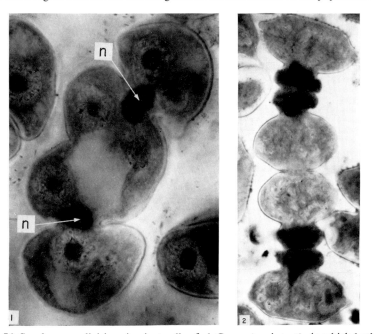

Fig. 54. Synchronous divisions in giant cells of: 1 *Cosmarium botrytis*, in which both nuclei are in very early prophase; 2 *Xanthidium antilopaeum*. Septum formation complete in each semicell. The two outer cells will develop into two normal semicells. n = nucleus. (Photomicrographs by Dr P. E. Brandham.)

of septum formation can give rise to a chain of incompletely separated cells. Brandham (1965) records a chain of 6 such double cells.

Although of very abnormal morphology, binucleate giant cells can divide quite normally and nuclear division and cytokinesis are synchronous at each end of the chain (Fig. 54). This occurrence has been reported by numerous investigators such as Ducellier (1915), Kallio (1957, 1961), West and West (1898) and Brandham (1964). Their formation and division in *Micrasterias americana* has been described in some detail by Rosenberg (1940) and Maezawa and Ueda (1979). They are also capable of conjugating with normal cells (Brandham 1964) (see p. 148) (see also Tan and Ueda 1978).

CHAPTER 6
SEXUAL REPRODUCTION
(CONJUGATION)

INTRODUCTION

Almost all students of biology have at some time observed, from photo-micrographs and drawings in textbooks, the major events in the process of conjugation by which the filamentous alga *Spirogyra* reproduces sexually. This mode of reproduction typifies the class Zygnemaphyceae and occurs not only in the family Zygnemataceae but in all desmids, both saccoderm and placoderm.

I. CONJUGATION IN SACCODERM DESMIDS

Sexual reproduction is seen with remarkable frequency in naturally occurring populations of many saccoderms, expecially in species of *Cylindrocystis* and *Mesotaenium* which form gelatinous strata in damp terrestrial habitats. It has also been observed in all other genera of the Mesotaeniaceae with the exception of *Ancylonema* (Fritsch 1935). Biebel (1973) reviewed the events of conjugation in 5 of the 6 genera of the Mesotaeniaceae all of which have been grown, and induced to conjugate in culture.

MESOTAENIUM

In the case of *Mesotaenium kramstai*, whose sexual reproduction was first described by Starr and Rayburn (1964), Biebel has induced conjugation in young cultures by transferring mixtures of two clones into N-deficient Bold's Basal Agar medium. At the onset of conjugation, which is essentially isogamous, potential conjugants pair up, sometimes with their longitudinal axes parallel, though orientation may be more or less at random (Fig. 55: 1, 5, 6). From adjacent walls, small fusion projections are produced which eventually develop to form a conjugation tube. The projections may grow out from almost any point of the cell wall, even the end (Fig. 55: 6), depending on the relative positions of the paired cells.

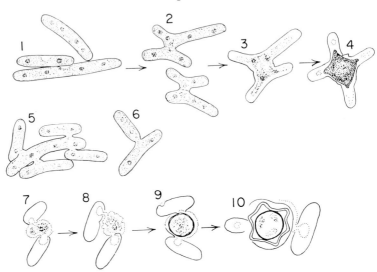

Fig. 55. Sexual reproduction in the saccoderm desmid *Mesotaenium*. 1–6 Stages in conjugation and zygospore formation in *M. kramstai* Lemm. (after Starr and Rayburn). 7–10 Stages in conjugation and zygospore formation in *M. dodekahedron* Geitler (after Geitler).

By solution, presumably enzymatic, of adjacent walls between the cells, a continuous tube is formed and into this the somewhat contracted protoplasts of each cell move, meet, and fuse (Fig. 55: 2, 3). At the same time the conjugation tube widens markedly so that it becomes increasingly less clearly demarcated from the walls of the original cells. Following protoplast (gamete) fusion, the resulting zygote develops a thick wall of several layers and remains for a time surrounded by the joined walls of the two fused protoplasts (Fig. 55: 4). The resulting zygospore becomes mahoganybrown in colour. Geitler (1965) gives a detailed account of conjugation and zygospore germination in *Mesotaenium dodekahedron* which differs in certain respects from other *Mesotaenium* species (Fig. 55: 7–10).

ROYA

Conjugation in *Roya obtusa* has been studied in some detail by Kies (1967) the process being easily induced in a homothallic clone. Papillae form adjacent to each other on pairing cells which remain some distance apart, enclosed in a structureless mucilaginous matrix secreted by the cells. The protoplasts round up and act as naked gametes, and these migrate from their enclosing cell walls through the papillae and fuse between them. Thus, no conjugation tube as such is formed, but rather a mucilaginous conjugation vesicle. The resulting zygospore develops a thick wall, but inside it would seem that the gamete nuclei remain unfused throughout a long period of dormancy.

Sexual reproduction in *Spirotaenia condensata* has been described by Hoshaw and Hilton (1966), and for *S. trabecula* by Biebel (1973). *Spirotaenia* differs significantly from all other desmids in that nothing like a conjugation tube or even a vesicle is produced. Like many other saccoderms the conjugating cells produce copious amounts of mucilage. The contents of the pairing cells, which consistently are significantly different in size, divide prior to fusion, the smaller of the pair doing so before the larger. Contractile vacuoles form in each of the daughter cells which shrink to become functional gametes. Hilton (1970) states that these show amoeboid movement and can change their orientation with respect to the larger cell. Non-sister gametes fuse following the gradual, but complete, gelatinization of the paired cell walls so that the final product is a pair of zygotes contained in a mass of mucilage. The zygotes develop characteristic alveolar wall thickenings, lose their chlorophyll, form one or more oil droplets and become separate zygospores as they enter a period of dormancy. Spontaneous germination of the zygospores has been observed some 3 weeks after their formation, and, in marked contrast to *Roya*, karyogamy occurs 3 days after plasmogamy. Meiosis is initiated at the onset of zygospore germination and the products are four germlings, which form within the zygospore wall. As these increase in length they are eventually released and develop into new *Spirotaenia* cells.

Of special interest is the fate of the chloroplast during the sexual cycle of *Spirotaenia* since it resembles, in some respects, that of *Spirogyra* and *Sirogonium* of the Zygnemataceae. As stated previously, the cells of *S. condensata* possess a single spiral chloroplast (see Fig. 22). The chloroplasts contributed by the 'male' gamete in *Spirogyra* disintegrate in the zygospore (Tröndle 1907, Allen 1958). In *Sirogonium* the chloroplasts of both male and female gametes fragment during gametogenesis, and the fragments re-organise in newly developing germlings. Although chloroplast fragmentation does not occur during gametogenesis in *Spirotaenia condensata*, only small fragments of the chloroplasts are found in the zygospore (Hoshaw and Hilton 1966). Newly released gones contain only a small amount of chloroplast material, and have no organised chloroplasts such as are found in germlings of *Spirogyra*. Hoshaw and Hilton suggest that comparative studies of chloroplasts of species of *Spirogyra*, *Sirogonium* and *Spirotaenia* might throw some light on the presumed evolutionary relationships of these members of the Zygnemaphyceae.

CYLINDROCYSTIS

Amongst the earliest observations of sexual reproduction in *Cylindro-cystis* were those in naturally occurring populations (Kaufmann 1914), and later in cultured material (Pringsheim 1918). The latter induced conjugation in *C. crassa* by omitting a nitrogen source from the culture medium. Biebel (1973) reports successful conjugation in cultures of *C. crassa*, *C. debaryi* and *C. brebissonii*; also in N-deficient media, at low light intensities (1000 lux). Although the different species showed differences in their time of response after being placed in the N-deficient medium, they all eventually did so. Cells paired and produced conjugation papillae at more or less any location on the cell wall where they came into contact. The subsequent events are very similar to those described above (p. 125) for *Mesotaenium*. The zygospores produced by the various species were found to differ somewhat in size and shape and in the extent to which the wall is thickened (Fig. 56: 1–3).

Biebel (1975) describes zygospore germination of a heterothallic strain of *C. brebissonii*. He stresses how the four plastids from the two original pairs of chloroplasts occupy the quadrants of the quadrate zygospore (Fig. 56: 4, 5) and that, although these gradually lose their

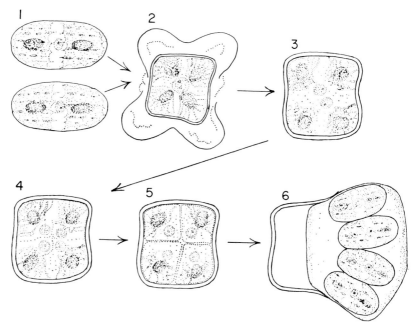

Fig. 56. Stages in conjugation (1 and 2) and zygospore germination (3–6) in *Cylindrocystis brebissonii* Menegh. (after Biebel).

green colour, the pyrenoid region can be identified throughout the entire period of dormancy. The first noticeable event marking the onset of germination is a change in the orientation of the plastids. Throughout dormancy they occupy the large horizontal plane, but at the initiation of germination, one pair comes to lie above the other in the narrow vertical plane. The protoplast divides by successive bipartition into four potential cells inside the zygospore wall. As they enlarge the spore wall ruptures and so they are released (Fig. 56: 6).

NETRIUM

A very complete account of sexual reproduction in two varieties of *Netrium digitus*, one homothallic, the other heterothallic, has been provided by Biebel (1964). Early accounts from natural populations are those of Pothoff (1928) and Grönblad (1957). Biebel induced conjugation quite readily by omitting a nitrogen source from modified Bristol's medium, when cultures were subjected to low light intensities at 20°C. In the case of the homothallic var. *digitus*, signs of conjugation were noticed within 48 hours. Sister cells of recent divisions can pair and a broad conjugation tube forms between them into which the shrinking protoplasts move and fuse with one another (Fig. 57: 1–3). A golden-brown zygospore is formed

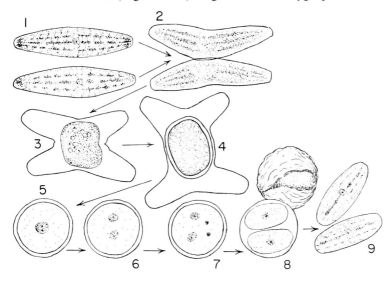

Fig. 57. Conjugation and zygospore germination in *Netrium digitus* (Ehr.) Itz. and Roth. 1–4 Stages in conjugation and zygospore formation. 5–9 Meiosis, the abortion of two of the resulting nuclei, and release of two germlings from the ruptured zygospore.

after a few days (Fig. 57: 4). Within this the chloroplasts disappear and the two gametic nuclei remain separate during dormancy.

Biebel records that after two months in a soil–water medium the zygospores so produced germinate spontaneously. During this process the chloroplasts reappear and the protoplast swells to rupture the surrounding outer spore wall, but it is retained within a vesicle. As these events are occurring, meiosis takes place, though of the four nuclei produced only two remain viable (Fig. 57: 5–7). Germlings, usually 2, are produced within the somewhat rigid vesicle membrane (Fig. 57: 9), though divisions of the nuclei may result in the production of up to 5 new desmids. In the aborting of two of the meiotically produced nuclei, *Netrium*, in this respect shows a characteristic of placoderm desmids (Fritsch 1935).

Biebel (1964) has shown that the heterothallic strain of the var. *lamellosum* of *N. digitus* shows some slight differences in its reproductive events. Thus when the two mating types from vigorous cultures growing on agar are brought together in N-deficient media they conjugate in 6 days. Their conjugation tubes are more bulbous than those of the var. *digitus* and as the protoplasts shrink contractile vacuoles, which fuse and discharge, appear on their surfaces, particularly in the region of the conjugation tubes. The mature zygospores, which are somewhat elongate, germinate in a fashion similar to those of var. *digitus*, though never more than 2 new cells have been observed to be produced in var. *lamellosum*.

II. CONJUGATION IN PLACODERM DESMIDS

Conjugation in placoderm desmids tends to be of much less frequent occurrence than in the saccoderms and especially in many planktonic species, it has yet to be observed. It seems to occur most abundantly in motile benthic or metaphytonic forms (p. 197) living in dense aggregations amongst aquatic macrophytes and their associated algal felts, especially in bogs and small pools (Coesel 1974). The most detailed accounts of it relate to various species of *Closterium*, several species of which can be made to conjugate in culture (Brandham 1964, Cook 1963, Coesel and Teixeira 1974, Dubois-Tylski 1972, 1973, 1975, Dubois-Tylski and Lacoste 1970, Lippert 1967, 1973, Pickett-Heaps and Marchant 1971).

Sexuality in desmids is either of a homothallic or heterothallic type, though both Brandham (1964) and Starr (1959) agree that this apparently simple state of affairs can sometimes be complicated by the existence of completely incompatible pairs of mating types within a species, but there is no evidence of 'relative sexuality'. This phenomenon has been recorded in the order Zygnemataceae of the Zygnemaphyceae, Hartmann (1955) having observed 3 conjugating filaments of *Spirogyra* in which the central one was behaving as a male with respect to one of the outer ones and as a female with respect to the other (see however p. 147 and Brandham 1967).

Brandham (1964) managed to isolate fertile cultures of 9 species of placoderm desmids from widely scattered localities in Britain. These are set out in Table 6.

Table 6. Fertile cultures of 9 species of placoderm desmids found in Britain.

Species	Mating type
Closterium ehrenbergii	heterothallic (6 out of 17 clones isolated were fertile)
C. leibleinii	homothallic
C. acerosum	homothallic
C. siliqua	homothallic
C. kützingii	homothallic
Cosmarium botrytis	mostly homothallic, some heterothallic (of 47 clones isolated only one failed to conjugate (see Table 7)).
C. subcucumis	heterothallic
Staurastrum denticulatum	homothallic
S. polymorphum	homothallic (14 clones from Epping Forest isolated and all fertile; 2 clones from Piltdown fertile. Not possible to cross clones from the two localities, so Brandham suggests obligate conjugation of sister cells.)

With regard to *Cosmarium botrytis* it can be seen from Table 7 that in many localities only one mating type is present and the complementary mating type is absent. This is particularly evident at Piltdown from which 11 + clones were isolated and no − clones; also at Chobham where the complementary mating type − is probably not present. In *Closterium ehrenbergii* Brandham found localities with only one mating type. For example, from a concrete horse trough in a farmyard at Maresfield, Sussex, he isolated 5 + clones, but no − clones. This he suggests, may at least in part account for the rare occurrence of conjugation of most placoderms in nature.

A review of the early literature pertinent to conjugation in *Closterium* is given by Lippert (1967) in an introduction to his well illustrated account of the process in *C. moniliferum* and *C. ehrenbergii* from which he isolated numerous strains from samples, including dried mud, taken from the margins of ditches, bogs, ponds and lakes. The most successful method for inducing conjugation in placoderms is based on that developed by

Table 7. Distribution of mating types of *Cosmarium botrytis* in nine localities in Britain (after Brandham 1964).

Locality	Variety	Mating Type+	Mating Type−	Mating Type+	Mating Type−	Homo-thallic	Sterile
Chobham, Surrey	—	2	2			6	
Gt Cumbrae, Scotland	—		1				
Dungeness, Kent	*subtumidum* Wittr.		1	4			1
Epping, Essex	—	1					
Elstree, Middlesex	—	3					
Goat Fell, I. of Arran	—	1					
Piltdown, Sussex	*tumidum* Wolle	11				3	
Radlett, Herts.	—	2	7				
Snowdon, Caerns.	*emarginatum* Hansg.	2					

Starr (1955) and in essentials consists of placing the strains to be mated in a watch glass within a Petri dish in which the surrounding air is CO_2-enriched by a 5% solution of $NaHCO_3$. A successfully used growth medium is Pringsheim's soil–water, deficient in an N-source. The cultures should be maintained at 20°C and illuminated for 18 hours in 24 at about 5000 lx. Dubois-Tylski and Lacoste (1970), whilst stressing the importance of an enhanced CO_2 supply to induce conjugation, did not find it essential for the nutrient medium to be N-deficient. Also, an increased CO_2 supply did not seem to be a prerequisite for conjugation in *Closterium littorale* (Pickett-Heaps and Marchant 1971). A further difference in induction technique comes from the experience of Dubois-Tylski (1973) with *C. rostratum*, which conjugated most successfully on agar slopes when provided with continuous low intensity illumination (1000 lux) using 'grow-lux' lamps. As pointed out by Coesel (1974a), conjugation would seem to occur most readily when illumination, temperature and CO_2 tension are at an optimum for mitotic activity. In nature, conjugation is most frequent in very shallow waters which can absorb CO_2 from the atmosphere; or possibly it is supplied by respiration from bottom muds. Such conditions prevail in bogs and temporary pools with changing water levels and where, clearly, the production of resistant zygospores will have considerable survival value (see p. 221).

It should be noted in this connection that Beijerink (1926) recorded that there is often a concurrent high incidence of sexual reproduction amongst representatives of quite distinct algal taxa (e.g. Oedogoniales, Zygnemataceae or Desmidiaceae) in a particular locality. Moreover, it can very frequently be observed to occur, at more or less the same time, in ecologically quite different, and far-removed, localities.

Coesel (1974) expresses the view that the scarcity, or complete absence, of observed zygospore formation in a large number of desmid species in nature can be attributable, in large measure, to a genetically determined incompatibility (see, however, Starr 1959). Coesel (1974b) isolated over 120 randomly selected strains of placoderm desmids belonging to 16 genera and more than 80 species and attempted to induce conjugation. Only 3 strains, all homothallic, exhibited conjugation. In discussing Starr's view (Starr 1959), Coesel points out that Starr 'does not seem to take into account that the strains he investigated were all isolated from natural populations in which there was already evidence of sexual reproduction, because he knew that randomly isolated strains cannot be easily induced to start conjugating. The success of his (Starr's) experiments is to a large extent attributable to the fact that his test material had previously already been selected for its sexual potential!' (see also Tables 6 and 7).

1. CONJUGATION IN CLOSTERIUM SPECIES

There are two somewhat distinct types of conjugation in the genus *Closterium*. The first type, which would appear to be the most common, takes place between two mature cells; the second involves the conjugation of two immature daughter cells which have just completed cell division.

Type 1

An example of the first type is provided by *C. rostratum* in which the stages in conjugation and zygospore formation have been described by Dubois-Tylski (1973) (Fig. 58: 1–6). As is frequently the case, the indications of imminent conjugation are manifest by an accumulation of starch and fat droplets in all of the cells. Two adjacent cells, which though fully formed have fairly recently divided, come together as a consequence of some force of attraction (see p. 147) and produce conjugation papillae from the middle of each lined-up pair of cells—presumably at a point along the line of the division suture (Fig. 58: 1, 2). There is some secretion of mucilage by the cells which, however, does not, as in so many other desmids, hold the cells tightly together. The papillae may come into contact soon after their inception, though frequently the cells are separated by as much as 50 μm, in which case the papillae elongate towards one another. The tips of the papillae are strongly vacuolated and active cytoplasmic streaming is apparent; as they expand the semicells open up at the suture line so that they appear geniculate (Fig. 58: 2).

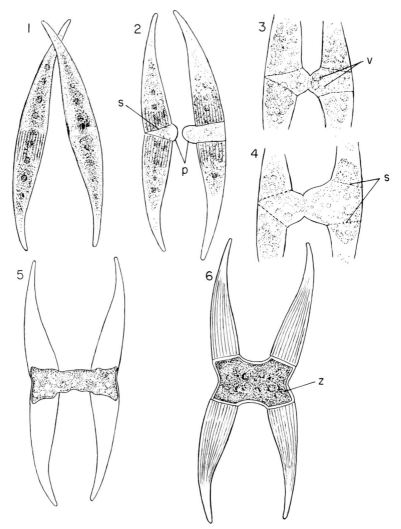

Fig. 58. Conjugation in *Closterium:* the *C. rostratum* type. 1 Initial pairing of cells; 2 protrusion of papillae as semicells separate along suture line; 3 papillae in contact, showing markedly vacuolated tips; 4 contraction and fusion of protoplasts in conjugation tube; 5 early stage in zygospore development showing its partial containment within the semicells; 6 mature, markedly angular zygospore. p = papillae; s = suture lines; v = vacuoles; z = mature zygospore.

The elongating papillae become a fairly well defined conjugation tube as they come into contact and join together by the dissolution of their adjacent walls (Fig. 58: 5). The protoplasts within the mother cells contract slowly (8–12 hours) and their eventual fusion occurs in the conjugation tube (Fig. 58: 4). Contraction of the protoplasts is rarely completed within the conjugation tube, but if it is the result is the formation of an elliptical zygospore. Most frequently part of the protoplast remains in the original pairing cells with the result that the zygospore is angular and partially contained within the semicells (Fig. 58: 5, 6). This angular type of zygote is considered typical of *Closterium* species of the *setaceum–kuetzingii* group.

In many other *Closterium* species, as for example the small types such as *C. parvulum* which have been studied in detail by Cook (1963), the cells of opposite mating types pair by the close association of the mid-region of the cells. A pore, often quite small in diameter, forms at the junction of adjoining semicells (Fig. 59: 1, 3) and the contents of each cell move into a conjugation papilla which slowly is extruded from the pore. An elastic wall, the investment, which appears to be continuous with the inner layer of the cell wall, surrounds the papilla (cf. Kies 1975, Abb. 1, p. 146). The enlarging papillae of the conjugating cells eventually meet and fuse to form a conjugation tube, within which fusion of the conjugant protoplasts takes place. Following this fusion, the resulting zygote decreases significantly in volume and becomes spherical. At the same time there is a deposition of a gelatinous matrix between the zygote and the wall of the conjugation tube so that, at maturity, the former is well separated from the enclosing wall. As Cook points out, this delicate conjugation wall, which can be stained with methylene blue, is persistent, so that the zygospore remains, until germination, in association with the parent cells from which it originated (Fig. 59).

Cook (1963) describes a new small species of *Closterium*, *C. evesiculum*, whose characteristic features are that although a conjugation tube is present in the early phases of sexual reproduction in the form of a gelatinous layer, this soon disintegrates. In consequence of the evanescent nature of its conjugation tube, the resulting zygote is easily dissociated from the parental cells.

Type 2

In *C. moniliferum*, *C. ehrenbergii* as well as in *C. acerosum*, *C. lineatum* var. *costatum*, *C. lunula* and *C. siliqua*, zygospore formation results from the conjugation of two immature daughter cells which have only just completed division (Lippert 1973) (Fig. 60: 1–5). Lippert (1967) found, when studying

K

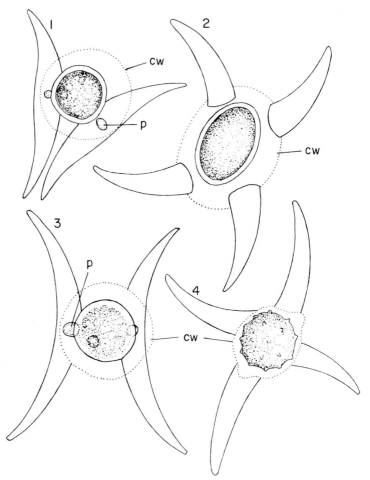

Fig. 59. Conjugation in small species of *Closterium* of the *parvulum* type. 1 *C. parvulum* Näg. var. *parvulum* Cook (after Cook); 2 *C. parvulum* Näg. (after West and West); 3 *C. dianae* Ehr. 4 *C. calosporum* Wittr. Note the empty cells are held together, with the zygotes between them, by the delicate conjugation wall (cw). p = the conjugation pore.

C. moniliferum, that cells about to conjugate showed the initial signs after the onset of darkness when on a 16-hour-light 8-hour-dark regime. Papillae, which are not always formed synchronously, appear soon after cytokinesis and arise from the single wall-layer of the small, conically shaped, immature semicells. The developing ends of these semicells usually lie adjacent to one another with mature semicells pointing away from one another, though they can lie side by side (Fig. 60: 1). Only occasionally

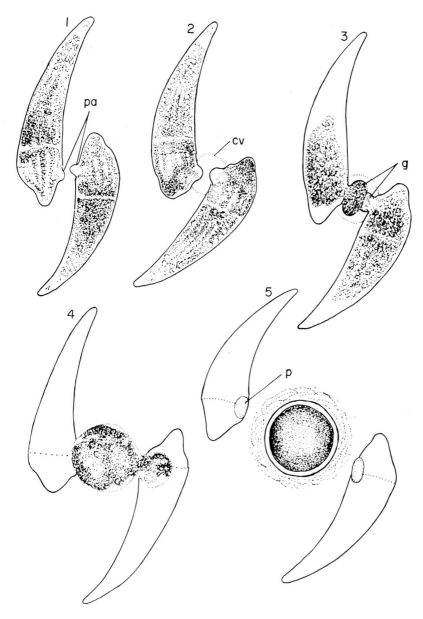

Fig. 60. Conjugation in *Closterium:* the *C. moniliferum–ehrenbergii* type. 1 The pairing of immature daughter cells and formation of conjugation papillae; 2 secretion of hyaline conjugation vesicle; 3 initiation of movement of gametes; 4 final stages of gamete fusion; 5 mature zygospore and empty cells showing conjugation pore. cv = conjugation vesicle; g = gametes; p = conjugation pore; p = papillae.

does pairing occur between non-daughter cells. A somewhat ephemeral, gelatinous conjugation vesicle is secreted and this appears as an enlarging hyaline bladder surrounding the two protruding papillae. Some 1–2 hours after the initial cell division before conjugation, the protoplasts begin to retract from the cell wall, especially towards the apices of the older semicells. Full gametic protoplast movement is preceded by the dissolution of the juncture between the two conjugation papillae and, as beautifully demonstrated in time-lapse films by Brandham (1967b), its initiation is sudden, though its progress is not always synchronous between pairs of cells, for the contents of one semicell usually move into the conjugation vesicle before the other (Fig. 60: 3, 4). Newly formed zygospores may at first be irregularly shaped, but they round up within an hour or so of their formation, and the twin zygotes remain between the empty cell walls of the gametangia (Fig. 61) held by the now considerably enlarged conjugation papillae. Maturation of the zygotes takes from 4–5 days

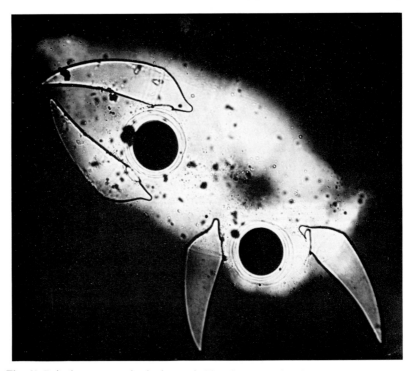

Fig. 61. Paired zygospores in the heterothallic *Closterium ehrenbergii* Menegh. produced from a cross between a + and − clone. Note the greatly enlarged conjugation papillae in the upper pair of conjugating cells; also the layers of mucilage surrounding the mature zygospores. The zygospores have been immersed in Indian ink to display the mucilage secreted by the conjugants (photo by P. E. Brandham).

during which time, as their walls thicken (Fig. 60: 5), the chlorophyll in the protoplasts disappears, and two dense, persistent plastids can be recognised. Although nuclear events were not observed by Lippert, Klebhan (1891) reported the presence of 2 nuclei in the newly formed zygospores of *C. lunula.*

Double zygospores have been described in some detail by Taft (1978) in *C. lineatum.* In their mature form these are ovoid with slightly produced angles at their opposed surfaces, which are flat, or more often, slightly retuse. Also they are remarkable in that they have a sharply defined and often wide-open suture in the median spore wall. Such sutures have not been reported in any other desmid, though they are present in all known species of *Spirogyra.*

There seems to be no clear pattern to the breaking of dormancy leading to the germination of the zygospores, though consistent germination of one clone was obtained by leaving freshly formed spores in the original mating dish, or by adding fresh medium to a microscope-slide preparation for several days.

Chlorophyll production marks the onset of germination and gradually the two persistent plastids 'green-up' and enlarge. At the same time the cytoplasm becomes filled with large granules and oil droplets. An event unique to the placoderm desmids (amongst the Chlorophyta) is the forceful release of a thin-walled vesicle containing the protoplasts through a fissure in the zygote wall. This seems to be a consequence of a rapid uptake of water in the early stages of germination, though once released the diameter of the vesicle diminishes noticeably, presumably through water loss. As a result, the developing plastids are brought into intimate contact. Although a fusion nucleus has not been observed, it is assumed that karyogamy takes place shortly before meiosis, since the first meiotic division takes place before the release of the vesicle. Cleavage takes place before the second division and one nucleus in the newly formed protoplast degenerates. Following cytoplasmic cleavage, the protoplasts begin to appear lunate and overlap, so that they remain closely associated as they enlarge and break through the thin vesicular membrane.

Details of the cytological events associated with conjugation at both the light and particularly the ultrastructural level have been followed in *Closterium littorale,* a species in which mature cells pair, by Pickett-Heaps and Fowke (1971). Paired cells become closely adpressed, and in each, a papilla is initiated as a narrow, circular band of diffuse wall material, quite different in appearance from the semicell wall. The organelle involved in papilla formation is stated to be the papilla vacuole, the enlargement of which, presumably by changes in turgor pressure, cause the localized growth of the papilla. The fibres of the two papillae were found to be

closely intermingled in areas of mutual contact and they grow asymmetrically, so that as the papillae balloon outwards, the area of wall in contact
increases considerably. The papillae themselves are formed by a rapid
and considerable phase of localized wall deposition and microtubules
appear adjacent to the newly formed papilla wall. As they develop the
cytoplasm begins to shrink away from the walls at the tips of each semicell
and the terminal vacuoles, so characteristic of *Closterium* cells, collapse
and disappear.

The expanding papillae become very large and flatten against one
another while they balloon out sideways. The papillae wall material
separating the cells by now is markedly thinner, but elsewhere it remains
quite thick. Although Golgi bodies, whose role is uncertain, remain
active during this period, the microtubules, which initially lined the
papillae, disappear and instead are now found at the edge, adjacent
to the old semicell. During this period a profound change occurs in the
general appearance of the cytoplasm, there being a considerable loss in
ground cytoplasmic structure. Although many ribosomes are visible, few
discrete organelles remain, apart from the nucleus and chloroplasts. In
the latter, the grana become increasingly distended and bloated.

Before protoplast fusion, the two protoplasts form a roughly spherical
mass within the distended papillae, though there are sometimes short
protrusions of cytoplasm still within the, for the most part, vacated
semicells. The papilla wall interposed between the two protoplasts has
by this time disappeared entirely leaving the two plasmalemmae very close
to each other. The actual fusion of the protoplasts is believed to be very
rapid but it is difficult to demonstrate. After fusion, the now binucleate
protoplast rounds up to fill, what could now be called, the conjugation
tube.

During maturation of the protoplast, the cytoplasm gradually regains
its granular and osmiophilic properties and the grana lose their bloated
appearance and become densely packed 'resembling the crystal-like grana
in the zygotes of *Chlamydomonas*'. Soon, however, the zygotes become so
dense that further ultrastructural observations become impossible.

Maturing zygotes which were sectioned by Pickett-Heaps and Fowke
(1971) all appeared to have two nuclei and they had no reason to believe
that their fusion was imminent, thus agreeing with nuclear events reported
in other instances of zygospore development. The shrinking zygote was
found to secrete a series of 6, thin-walled layers, and a brief phase of Golgi
activity seemed to accompany the formation of the first few. It was doubted
whether Golgi activity was involved during the long development of the
main stratified wall.

2. CONJUGATION IN PLEUROTAENIUM

Conjugation has been described for two heterothallic strains of *Pleurotaenium*, *P. ehrenbergii* and *P. trabecula* var. *mediolaeve* by Ling and Tyler (1972a). Starr's method of inducing conjugation was successfully applied and, as found with other desmids, it occurred first at the edges of the watch glass in which the clones were mixed and spread centripetally, presumably in response to a CO_2 gradient.

Conjugant cells lie parallel to one another with their isthmuses adjacent and an enveloping secretion of mucilage holds the paired cells together. Soon a mucilaginous conjugation vesicle is secreted, presumably from mucilage pores of all four semicells in their isthmal regions; and this pushes the cells apart (Fig. 62: 1). The isthmuses lengthen but the semicells remain joined by a thin new wall. At the same time, the protoplasts condense, drawing away from the end walls.

At the isthmus of one cell a hyaline papilla protrudes into the conjugation vesicle, towards the other cell. This papilla increases in size as the protoplast is extruded from the cell into it; the other cell also produces a papilla, and when both are about 6 µm long the cells apparently become quiescent for 1–2 hours (Fig. 62: 2). Then quite rapidly, the first-formed papilla enlarges followed by the second until they touch. The membranes separating then dissolve allowing their contents to move together and fuse. One cell may empty before the other, though not necessarily the one producing the first papilla (Fig. 62: 3).

As the zygospore starts to mature the conjugation vesicle enlarges, but before the zygospore walls are laid down, the zygote shrinks considerably. During maturation, 4 or 5 zygospore wall layers are laid down (Fig. 62: 4). The first is thin and membranous and is followed by a thick, hyaline layer immediately beneath it (exospores I and II), and this is finely punctate like the parent cell wall. A middle or mesospore wall is mammillate and it is this which turns brown as the zygospore matures, while the innermost layers (endospores I and II) are transparent.

The mature zygospores are laterally compressed and lie between the empty cells, the flattened faces lying adjacent to these mother cells. A remarkable, and apparently unique, feature of the *Pleurotaenium* zygospore was observed when some were compressed under a coverslip. An ellipsoidal plate which had been located on one of the flattened sides was detached from exospore II. As Ling and Tyler (1972a) point out, if this operculum is not found in other desmids, it may prove to be a useful generic criterion. However, Blackburn and Tyler (1980) have found in a very recent study of conjugation, and subsequent zygospore germination in *Micrasterias mahabuleshwarensis*, that the germination vesicle also escapes following the opening of a hinged operculum of the exospore, which bears 2–3 fork-tipped spines.

Germination of *Pleurotaenium* zygospores, and, as a consequence, the
part played by the operculum described above, have been followed by
Ling and Tyler (1972b). They found the period for dormancy for *P.
trabecula* var. *mediolaeve* to be short and that spontaneous germination

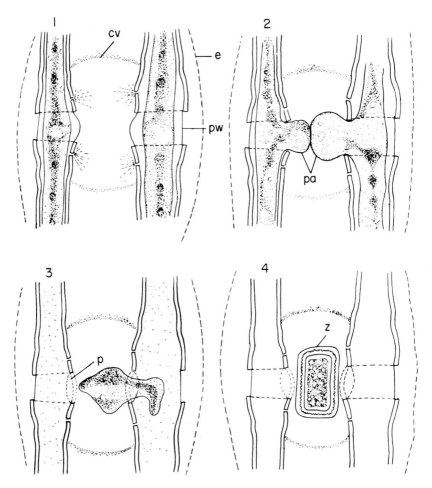

Fig. 62. Conjugation in *Pleurotaenium*. Diagram of the process of conjugation in
Pleurotaenium. 1 The elongation of the isthmus and secretion of the envelope (e)
which initially holds the paired cells together, and the conjugation vesicle (cv); 2 the
development of papillae (pa); 3 fusion of the gametes; 4 the maturation of the
zygospore (z). p = gamete exit pore, pw = the newly secreted wall. (After Ling and
Tyler.)

may occur within three weeks of their formation; for *P. ehrenbergii* it takes much longer. Germination can be seen to have started when the germination vesicle begins to squeeze outwards, rupturing the brown mesospore and forcing the operculum outwards. The operculum may be cast off or remain partially attached and so appear hinged. The protoplast, by now much larger than the zygospore, moves out though still enclosed by two endospore walls. Within it starch and oil globules can be seen to occupy the periphery of a large, clear central vacuole and from 1 to 3 dark green chloroplasts may be visible. As in the case of *Closterium littorale* (see p. 139) the swelling of the protoplast suggests that the absorption of water initiates the early events of germination. Later it shrinks and then both walls of the surrounding endospore become visible. Some time later the protoplast becomes pinched by a furrow, into two unequal portions. This furrow is the sinus of the single product of germination and so may be the protodesmid, or gone (Starr 1955). As previously mentioned (p. 130) in the majority of placoderm desmids two gones usually result from meiosis in the zygospore but, as in *Pleurotaenium*, the survival of a single meiotically produced nucleus has also been found to occur in *Cosmarium biretum* (Starr 1959) and *Hyalotheca dissiliens* (Pothoff 1928).

The unequal proportioning of the gone in *P. trabecula* (it does not occur in *P. ehrenbergii*) has two effects on subsequent divisions. Firstly, the smaller 'heteromorph', with the smaller gonal semicell, divides more slowly than the larger one; and secondly, its daughter cells are usually smaller. This must clearly result in two distinct size classes in ensuing clones. To what extent this may occur elsewhere in natural populations is not known, but if it does, it seems to offer a possible explanation for the distinct size differences reported for some desmids.

3. CONJUGATION IN THE COSMARIAE

It is most interesting to note that Kies (1968) has found that conjugation in *Micrasterias papillifera* resembles that of *Pleurotaenium*, in that the pairing of conjugants which come to lie face to face is initiated with the production of a conjugation vesicle. This is believed also to originate from mucilage produced from pores adjacent to the isthmus. In a subsequent EM study of the process in *M. papillifera* (Kies 1975) it has been demonstrated that papilla formation is preceded by the intercalation of a newly formed strip of partially cellulosic wall at the isthmus between old and new semicells, and it is from this that the papilla originates. The papillae enlarge rapidly and their walls become very thin prior to fusion of the protoplasts, so that the plasmalemma of these potential gametes are almost in contact. Kies reports that there is no breakdown of the plasmalemma along their entire line of contact but rather the establishment of a centrally situated 'fusion-bridge' which enlarges centrifugally, pushing the remaining plasmalemmas, and the remnants of the papillae walls, to the outside. A film is available of conjugation in *M. papillifera* (Kies 1971).

Conjugation and zygospore formation have also been described for *Micrasterias rotata* by Lenzenweger (1973) and seem to be quite typical of the majority of the Cosmariae in which they have been studied. They have been followed in detail by Starr (1954) and Brandham (1966) in *Cosmarium botrytis*, the latter worker using time-lapse cinematography (Brandham 1967). They have also been described for a heterothallic strain of *Staurastrum gladiosium* by Winter and Biebel (1967). Brandham recognises 5 fairly distinct phases in the process of conjugation in a typical member of the Cosmariae:

1   Pairing, in which potential conjugants come together and are at first orientated at random with respect to each other.

2   Reorientation in which the conjugants move into a position so that their broadest faces are touching, and their longitudinal axes are at right angles to one another. During this phase the paired cells secrete an enveloping sphere of mucilage (Fig. 63: 1).

3   Formation of conjugation papillae which grow towards one another and while doing so each conjugant separates at its isthmus (Fig. 63: 2).

4   Gamete fusion which involves the migration by amoeboid movement of the protoplasts, now isogamous gametes of the paired cells (Fig. 63: 3). These fuse in the considerable space between the cells. After fusion the empty cell walls usually remain close to the zygote since they are embedded in a copious mucilage envelope (Fig. 63: 4). This is followed by a marked contraction of the zygote due to the discharge of numerous contractile vacuoles (Brandham 1967).

5   Zygospore formation and maturation involves the development of a 3 (5–6?) layered wall consisting of thin endospore, thick resistant mesospore and a water permeable exospore (Fig. 63: 5). The latter in very many species, becomes characteristically ornamented with warts and spines before the formation of the inner walls (Fig. 63: 6). In many cases it seems to be a specific character and in some major taxonomic keys (e.g. West and West 1901–1919 and Smith 1924) details of zygospore shape, size and ornamentation are given along with descriptions of the vegetative cells.

Fritsch (1935) stresses that before conjugation starts in filamentous desmids, there has usually been a dissociation of the filaments into individual cells, exceptions being some species of *Desmidium*. Except in *D.*

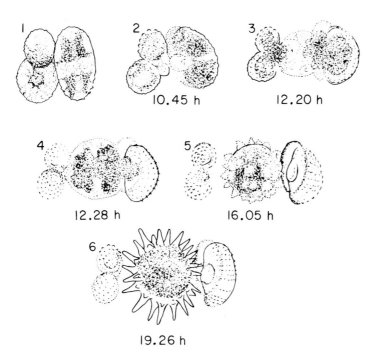

Fig. 63. Major events and their timing in the conjugation of *Cosmarium botrytis* Menegh. 1 Late stage in pairing with the longitudinal axes of cells orientated at right angles to one another; 2 protrusion of conjugation papillae pushing cells apart (10.45 h); 3 gametes emerging, but not yet fused (12.2i h); 4 fusion of gametes; one on left now completely emerged (12.28 h); 5 early stage in spine formation on zygote wall (note little contrast between spines and surrounding mucilage) (16.05 h); 6 spines almost fully developed. Their greater contrast, as compared with 5, suggests a change in the nature of the spine material (19.26 h). (After Brandham 1964.)

cylindricum where the zygospore is lodged within the female cell (Schulz 1930)—an oospore?—the zygospore is formed between the conjugating cells. Much more needs to be known, however, about the process in filamentous forms, since detailed knowledge is meagre.

In his time-lapse cinematographic study, Brandham (1967) determined the times taken for the various phases of conjugation to be completed. He also measured the speed of movement of cells. Although speeds of 100 µm per min were measured when cells were placed in fresh medium (Brandham 1965), this was reduced to 20 µm or less just before conjugation. He also (1967) analysed the anisogamous mating behaviour of 14 clones of *Cosmarium botrytis*, some of which were triradiate, instead of the normal biradiate shape (Fig. 64: 1, 2). In all crosses the minus cells (see p. 130) almost always approached the plus cells directly and there was no indica-

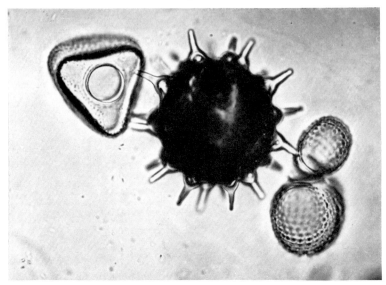

1

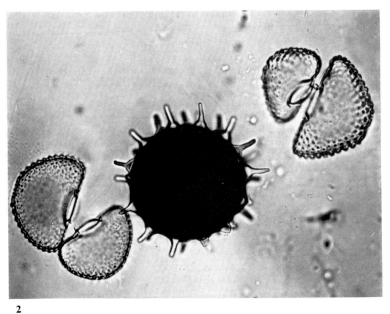

2

Fig. 64. Anisogamous mating in *Cosmarium botrytis* Menegh. 1 A zygospore produced by crossing a triradiate clone (left) with a biradiate clone (right); 2 zygospore production by the conjugation of different clones of the var. *emarginatum* Hansg. (S2 right and R2 left). Note that the 4 empty semicells are morphologically different. (Photomicrographs by Dr P. E. Brandham.)

tion of random meetings. Brandham suggests that almost certainly because of this behaviour, a chemotactic response is involved and that the 'minus' cells are moving along a diffusion gradient of a sexually attractive substance secreted by the 'plus' cells. He also showed that this so-called anisogamous mating is not rigidly linked with mating type since some 'plus' clones are either passive, or chemotactically active, depending on which 'minus' clones they are crossed with. This 3-group anisogamous mating behaviour is explained on the basis of a 2-part hormone secretion produced by passive cells, one of which is the precursor of the other, and less efficient in causing the chemotactic response. Production of the precursor and complete hormone are presumed to be controlled by separate genes (Brandham 1967, see Sasaki 1978).

Winter and Biebel (1967), in studying conjugation in *Staurastrum gladiosum*, in which no conjugation tube was formed, observed the gametes escaping into a common matrix of mucilage between the paired cells; one gamete emerged first and that one always flowed into the other during fusion. They comment that 'it is possible that one mating type consistently emerges first, or that one mating type gamete always flows into the other'. They were, however, unable to confirm this, but indicate that it could be by the use of a triradiate clone and a complementary clone with a different cell morphology (bi- or quadriradiate).

It is worthy of note that in Turner's (1922) observations on conjugation in *S. dickei* var. *parallelus*, the population was polymorphic with bi- and triradiate forms occurring most abundantly but with quadriradiate forms also present. Presumably because they were more numerous, conjugation was quite commonly seen between bi- and triradiate cells, though he reveals that he has 'several times seen conjugation of 3-rayed with 4-rayed forms, but has not yet observed the conjugation of two 4-rayed modifications though I have searched for them'.

When zygotes of *Staurodesmus dickei* germinated Turner noted that 'three-rayed desmids sometimes occur in the same protoplast with the four-rayed forms—that is to say, that the same spore gives rise to both kinds simultaneously'.

Brandham (1967) reports that many of the pairings of *Cosmarium botrytis* are temporary and that the cells disassociate after a few hours, a phenomenon which he has termed 'pseudo-conjugation'. It was noted that pseudo-conjugating cells never secrete the normally seen sphere of mucilage around themselves. Although various types of pseudo-conjugation are described, the commonest cause would seem to be the occurrence of vegetative division of one of the cells, usually the passive + cell. Brandham suggests that the attractive substance constantly secreted during normal conjugation ceases when the nucleus passes from the metabolically active interphase to the less active state associated with mitosis. Other causes invoked are a permanent or temporary cessation of the secretion of the chemotactic substance 'because of various factors as yet unknown'!

Of some significance in this respect would seem to be the behaviour during sexual reproduction of diploid cells (Fig. 65). Brandham (1964) produced diploid from haploid cells of *Cosmarium botrytis* and found their mating behaviour to be the same as that of the haploid clones from which they originated, being heterothallic and of minus mating type. He was able to cross these diploids successfully with haploids and found one very unusual feature about this phenomenon. The majority of such conjugations involved one diploid cell, flanked on either side by a haploid (Fig. 66: 1). The conjugation cells paired, isthmus to isthmus, with their longitudinal axes at right angles, and then the central diploid cell extruded conjugation papillae towards each of the two haploids. Sometimes one of the haploids failed to produce a papilla and so took no further part in the conjugation process.

Most frequently, however, the diploid withdrew one of the papillae and its contents fused with one of the haploid cells, while the other aborted leaving either a mass of starch granules or the developing gamete which

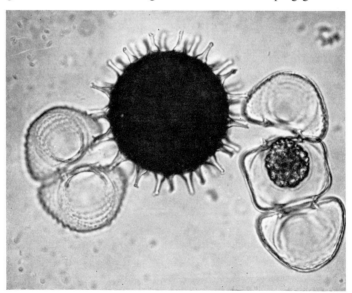

Fig. 65. Conjugation between a giant cell and a normal cell in *Cosmarium botrytis* var. *tumidum* Wolle fac. *triradiata*. The contents of the lower semicell of the giant cell has moved into the central rectangular, abnormal region, and has aborted following the successful conjugation of the contents of the normal cell and the combined contents of the upper semicell and adjoining giant cell. (Photomicrograph by Dr P. E. Brandham.)

divided into a parthenospore (see p. 155) (Fig. 66: 2). Much less frequently, the diploid gamete divide into two parts each of which fused with a haploid gamete. The resulting pair of zygospores were of equal size. The rarest behaviour of all, and which Brandham observed on only three occasions, was the fusion of the contents of all three gametes to form a zygospore containing one diploid and two haploid nuclei (Fig. 66: 3).

The giant binucleate cells, which are often encountered in old cultures, though somewhat rarely in nature (see p. 122), have been demonstrated by Brandham (1965) to retain their sexual activity and conjugate with normal cells. When this happens the giant cell dehisces only at one of its two isthmuses and its contents fuse with the protoplast of a normal cell to produce a trinucleate zygospore. Such an event may be complicated by the presence of a second normal cell which also attempts to conjugate with the giant cell. The contents of the second normal cell always abort when the gametes of the giant cell and the first normal cell fuse.

The diversity of types of normal conjugation encountered in the desmids has been reviewed by Ling and Tyler (1972) and Dubois-Tylski (1973). The two fundamental differences would seem to hinge on whether a conjugation tube or conjugation vesicle is formed. Thus the following categories seem to be distinguished.

1. *Conjugation by a conjugation tube*
This type is frequent amongst the saccoderms, and the tube formed by the extension of conjugation papillae consists of a simple cellulose wall which is in effect an extension of the walls of the pairing cells. The tube, which may become considerably distended, persists as part of the zygospore wall and helps in maintaining its solid shape. It occurs in *Netrium digitus*, *Mesotaenium kramstai* and *Cylindrocystis* spp. of the saccoderms and in the placoderms in *Closterium* species of the *parvulum*-type (Cook 1963).

2. *Conjugation through the mediation of a conjugation vesicle*
There are various types of conjugation vesicles which according to Dubois-Tylski (1973) may be separated into the following categories.
a The type described by Kies (1968, 1975) for *Micrasterias papillifera* in which conjugation occurs within a preformed, delimited mucilaginous vesicle.
b Conjugation vesicles without distinct limits as found in *Roya obtusa*, *Pleurotaenium* spp. and possibly *Closterium evesiculum*. Cook (1963a) describes it, however, in this species, as a mucilaginous conjugation tube without a definite wall and which does not persist.
c The conjugation vesicle which is secreted by the papillae themselves during their formation. This is not bounded by a membrane and it disappears almost at the actual moment of conjugation. It is typified by

Closterium moniliferum and *C. ehrenbergii* and has been incorrectly termed a tube by various authors.

d The Cosmariae type of vesicle which usually becomes extensive, and forms as the conjugants orientate themselves during the later stages of pairing. It occurs for example in many, if not all, *Cosmarium* and *Staurastrum* species, possibly in *Closterium littorale* and in the one saccoderm *Mesotaenium dodekahedron*.

e In *Spirotaenia condensata* the entire wall of each conjugating cell becomes gelatinised with the result that the protoplasts fuse in a structureless mucilaginous envelope.

As pointed out by Dubois-Tylski (1973b) the various modes of conjugation found throughout the desmids have little systematic significance. To what extent they have ecological significance in that they afford protection to the gametes during the course of sexual reproduction rather than

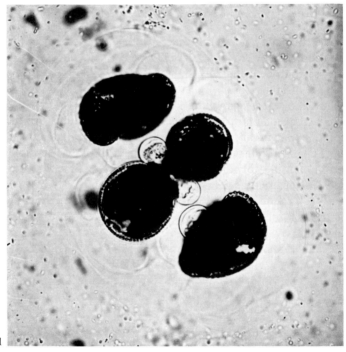

Fig. 66. The sexuality of diploids of *Cosmarium botrytis* Menegh. 1 a diploid cell (centre) conjugating with two haploids, showing two papillae produced by the diploid and one by one of the haploids. Note the copious mucilage which surrounds the conjugants. 2 A zygospore produced by the fusion of the diploid gamete with one of the haploids. The other haploid (a) has aborted. 3 A diploid cell has conjugated with two haploids, all three having fused to form a tetraploid zygospore. (Photomicrographs by Dr P. E. Brandham.)

tiple pairs which they observed suggests that *P. coronatum* produces a powerful sex attractant, and they regard the small cells of *P. ehrenbergii* and *P. mamillosum*, which conjugate around the *P. coronatum* cells, as males. This conclusion would seem to be supported by the fact that they observed a considerable number of aborted cells, and a precocious production of conjugation papillae.

Desmid taxonomists in the past (West and West 1904–12, West and Carter 1923) have tended to place considerable emphasis on the morphology of zygospores in the recognition of taxa. It is clear that zygospore spine morphology as a taxonomic criterion should be treated with caution, since their development can be considerably influenced by cell turgor (Coesel and Teixeira 1974, Ichimura and Watanabe 1974, Kies 1968, 1970).

ASEXUAL SPORES

There have been reports from time to time by algologists from various parts of the world of the asexual production by resting spores in a fairly wide range of desmid species. Almost all of the early reports seem to have been based on observations made of preserved material and so the interpretations as to how such spores were formed, and under what conditions, were inevitably largely speculative. Indeed Lhotsky (1948) thought that those which he observed in a species of *Closterium* were degenerate cells formed under adverse cultural conditions, while Huber-Pestalozzi (1927) interpreted those found by Bennet (1892) in *Closterium lanceolatum* as possibly a consequence of infection by some chytrid fungus.

Starr (1958) states that, with the exception of Nieuwland's (1909) account of resting spores in *Cosmarium bioculatum*, most accounts can be explained as a result of environmental conditions which cause plasmolysis of the desmid's cell contents, followed by the secretion of a resistant wall round the contracted protoplast. In *C. bioculatum*, Nieuwland describes how cells broke at the isthmus and discharged their entire contents into a mucilaginous mass which surrounded them. The discharged protoplast then became spherical and secreted a spiny wall, typical of the zygospores' characteristic of the species. Kniep (1928) has interpreted spore formation as a reduced type of sexual reproduction in which the two halves of a single cell fuse. He assumed that this was preceded by a nuclear division, thus providing each half cell with the nucleus necessary for the presumed fusion.

Starr's (1958) observations of the formation of asexual spores in *Closterium didymotocum* do not support Kniep's view (see however, Starr 1954) since he found no evidence of a nuclear division forming sister nuclei prior to spore formation. Starr describes how on finding this *Clos-*

terium, asexual spores were being formed fairly freely and that attempts to establish cultures of the desmid were unsuccessful. However, he collected soil from the area and allowed it to dry. Some 5–6 months later the air-dried soil was wetted with distilled water and illuminated. Within a week vegetative cells appeared. After a period of normal cell division, however, they began to form spores once more. Clones were isolated and grown in tubes in low light intensity (350 ft.c.) with a light regime of 16 hours light and 8 hours dark, in Waris' solution. Although there was once more an initial growth of vegetative cells, within a week spore production recommenced. Transfer to fresh medium prolonged the period of normal multiplication for a time so that quite a number of normal *Closterium* cells were produced but again, after a while, spores began to form.

As in Nieuwland's *Cosmarium bioculatum*, asexual spore formation in *Closterium* was accomplished by the disjunction of the cell wall, though, because there is no isthmus, it took place between an intercalary band and an end wall piece. Occasionally it happened at a suture between adjoining intercalary bands. A small band of mucilage was secreted at the point of disjunction prior to the emergence of the protoplast. As the latter emerged it formed a spherical mass lying between the two portions of empty cell wall. As this mass matured to become a dormant spore, it secreted a secondary resistant wall between the protoplast and the first bounding membrane. The chloroplasts then rapidly lost their chlorophyll and were no longer visible. Although some germination of these spores was observed, there were insufficient instances to permit cytological investigation of the nuclear events associated with it.

Lhotsky (1973) comments on the very rare occurrence of chlamydo-spores in nature and reports on finding them on only three occasions despite long searching. The species in which such spores were found were *Closterium juncidum* var. *brevis, C. striolatum* and *C. moniliferum*. Because they were found in very small bog pools, Lhotsky suggests that the population involved were clones of one of the mating types of hetero-thallic strains and that their only way of surviving unfavourable condition was by producing akinetes. However, as he points out, the real asexuality of these bodies 'has not been proved by genetic experiments up to now'.

Coesel and Teixeira (1974), in conjugation experiments with unialgal cultures of *Micrasterias papillifera*, observed on occasions, in sexually reproducing material, the formation of parthenospores. In such cases, the contents of a simple cell protruded from the isthmus rounded into a smooth globose cell, but failed to form the spines characteristic of this species' zygospores. Ling and Tyler (1972) record the formation of parthenospores in *Pleurotaenium ehrenbergii* cultures after a failure to complete conjugation successfully.

The most detailed study of haploid resistant spores carried out so far is that of Brandham (1965) in which he observed its occurrence in 3 species of *Closterium*, several strains of *Cosmarium botrytis*, *Pleurotaenium trabecula* and *Staurastrum denticulatum*. His study has suggested that 3 major types of spores may be produced depending on the relationship of spore formation with respect to the parent cell. These are:

1. *Emergent spores*

Spores, such as those found by Nieuwland (1909) in *Cosmarium bioculatum* and Starr (1958) in *Closterium didymotocum*, in which the cell breaks open at the isthmus in the case of *Cosmarium*, or at a suture line in *Closterium*. The cell contents emerge into an envelope of mucilage previously secreted by the cell, round up and form a resistant wall similar to that of their zygospores. Brandham suggests that since they are produced completely asexually, they may therefore be regarded as aplanospores. By contrast, in *C. kuetzingii*, they are produced in actively conjugating material and so possibly are parthenospores, the other conjugant having aborted. Brandham also records a pair of parthenospores resulting from the failure of gamete fusion in *Cosmarium botrytis*, *Closterium ehrenbergii* and *Staurastrum denticulatum*. Parthenospores were also observed in *Cosmarium botrytis* var. *tumidum*, and in this the spores and parental cell were embedded in a sphere of mucilage. No trace of the presumably aborted conjugant was observed, so it seems possible that these were in fact produced without any sexual stimulus even though conjugation was taking place in the culture. In the case of *Staurastrum denticulatum*, although conjugation and normal zygospore production were occurring, other spores were produced in which there was a complete absence, or only partial fusion, of pairs of cells. The resulting spores were of very variable shape.

2. *Semi-emergent spores*

These were seen only in *Cosmarium botrytis* by Brandham (1965), though previously reported for *Staurodesmus dickei* by Turner (1922). Brandham describes how during conjugation each conjugant splits at the isthmus and a conjugation papilla extends towards the partner's isthmus. One of the conjugants, however, aborts while at the papilla-formation stage. The protoplast of the surviving conjugant does not emerge, but remains in the cell with its extruded papilla still apparent. Initially a thin wall is formed around it, and later a thick 3-layered wall typical of a zygospore is secreted.

3. *Non-emergent spores*

This type of spore has been observed occasionally by Brandham (1965).

In these the protoplast does not contract but secretes a single-layered thick wall inside the original cell wall which thus differs from the walls of zygospores or parthenospores. They have been described for *Closterium lanceolatum* and *C. striolatum* (Bennett 1892), *Hyalotheca neglecta* (West and West 1898) and in *Tetmemorus laevis* and 3 *Cosmarium* species (Huber-Pestalozzi 1927). As Brandham points out, Klebs (1896) induced the formation of similar spores in *Closterium* and *Cosmarium* cells by plasmolysis. Brandham tried to repeat this technique with *Cosmarium botrytis*, but his plasmolysed, cells failed to secrete a wall.

ARTIFICIAL SEPARATION OF DESMID PROTOPLASTS

As stated above, it has been suggested that spore formation, since it involves the rounding off of the desmid protoplast, may be plasmolytically induced. Of considerable significance in this connection is Chardard's (1972) study of protoplast production in *Cosmarium lundelli* by plasmolysis. In his experiments he used two types of liquid culture media; that of Lefèvre supplemented with what he terms 'oligo-elements', and the other was Pyrex-glass distilled water. To each he added sucrose and mannitol in concentrations of 0.3, 0.4 and 0.5м. In the hypertonic Lefèvre media the cells died rapidly, he presumed as a result of an accumulation of mineral salts in the cells. However, in distilled water made hypertonic with mannitol or sucrose, there was quite a considerable survival, even though the cells became plasmolysed. Differences were observed in the way in which cells became plasmolysed, some protoplasts rounding up in one semicell, others forming somewhat irregular shapes, others occupying the space in the region of the isthmus with a portion of each protoplast in each semicell. 0.5м sucrose produced the greatest incidence of plasmolysis (43%) after 3 weeks exposure. With mannitol, although the proportion of cells plasmolysed was not as great, their survival was better than in sucrose. The optimal concentration was 0.4м.

An observation of special significance in relation to desmid spore formation is that when a protoplast rounds off in the middle of a cell at the isthmus, the resulting pressure exerted on the wall tends to separate the two semicells and the naked protoplast moves out. It is also interesting to note that in some cases anomalous cell divisions are observed, such as the incomplete formation of a new cell wall and the formation of giant cells (see p. 122), or the unequal partition of the protoplast between the two semicells. Of especial interest is the observation that occasionally protoplasts have been seen with two concentric walls, presumably secreted by the extruded protoplast. Chardard also noticed some protoplasts, smaller than others, with a delicate wall in intimate contact with the cytoplasm.

Similar observations to those of Chardard have been made by Berliner and Werc (1976a) with *Cosmarium turpinii*. They found that these desmids can be induced to release their protoplasts when incubated for 3–5 days in a mineral–soil–water medium, contrary to Chardard's experience, and 0.4M mannitol. At this concentration the protoplasts of each semicell rounded up and one passed across the isthmus into the other to form one large, dense, membrane-surrounded structure. Giant cells and triple cells, in addition to other morphological abnormalities indicative of abberations in cell division, were found only in mannitol concentrations of 0.6M. Berliner and Werc point out that the protoplasts are approximately the same size and shape as the zygospores of *C. turpinii* as described by Starr (1955). These investigators question whether, as Chardard (1972) suggests, protoplast release is a purely physical process, or due in addition to the activation of the cell's own autolytic enzymes.

In a subsequent paper these authors (Berliner and Werc 1976b) put forward the interesting proposition that, since *C. turpinii* protoplasts can be obtained so easily merely with mannitol treatment, they could be a valuable and readily produced source of unicellularly derived protein.

AKINETE FORMATION

According to Fritsch (1935), when cells undergo thickening of their membranes and at the same time accumulate reserves of oil and/or starch within their protoplasts, they can be said to have formed 'akinetes'. These structures can always be distinguished from aplanospores (and parthenospores) by the fact that the additional wall layers around the protoplast are fused with the wall of the parent cell (Smith 1950).

Lhotsky (1948) described a structure of this type in *Closterium moniliferum* from a culture in which conditions seemed to be unfavourable for growth. In their production, cells were formed in which one semicell appeared normal while the other was small and rounded. A transverse wall was formed half way along the larger semicell and the distal portion cast off. The remainder formed a thick wall. Brandham (1965) records a similar sort of akinete in a culture of *Pleurotaenium trabecula* in which growth had slowed down due to nutrient depletion. Newly produced semicells usually remained short, and in some, a stage was observed in which the entire contents had contracted into the smaller semicell. This then became separated by the formation of a cross wall. A similar type of cell, with one normal length and one short semicell, was found in a naturally occurring population and was successfully cultured. In this, the small semicell became very dark and produced a single very thick wall, subsequently separating from the larger semicell. The protoplast of the latter died.

CHAPTER 7
THE NUCLEAR CYTOLOGY
OF DESMIDS

Details of the nuclear organisation and mitotic behaviour in many algal groups are still far from completely known, so that no general statements about the organisation of the genetic material within any of these primitive plant groups can yet be made with confidence. However, it does seem that some quite fundamental differences exist between them, as demonstrated by studies of the Euglenophyta (Leedale 1968) and the Pyrrophyta (Dodge 1963). With regard to the Zygnemaphyceae, studies of *Spirogyra* (Geitler 1930, 1935, Godward 1950, 1953, 1954a, b, 1956, 1961, Godward and Newham 1965, Godward and Jordan 1965) and of numerous desmid genera (King 1953a, b) and especially of the genus *Micrasterias* (Kallio 1950, Kasprik 1975, Brandham 1964) would seem to set this class apart, not only from other Chlorophyta, but to some extent from the rest of the Plant Kingdom. The features of very special significance are the large size, and apparent complexity, of the interphase nuclei, the remarkable organisation and persistence of the nucleolar material, and the nature of the centromeres of the chromosomes. Thus in Godward's review of the Zygnemaphyceae (Conjugatophyceae) in '*The Chromosomes of Algae*' (Godward 1966), she focusses attention on the common (though not invariable) absence from the chromosomes of localised centromeres, and the presence during mitosis of stainable material derived from the nucleolus. This often occurs round the metaphase plate and is distributed with the anaphase movements of the chromosomes.

1. THE NUCLEOLUS

As pointed out by King (1959) our knowledge of the nuclear cytology of the desmids is still scanty, but all studies focus some attention at least on the nucleolus. Kopetzky-Rechtperg (1932) was the first to note that the number of nucleoli varies even between species within a given genus, some having single nucleoli; others many.

In one of the first detailed studies of nuclear division, Waris (1950) investigated its progress in a number of *Micrasterias* spp. which have multiple nucleoli. He found that cells ready to divide can usually be easily recognised by their enlarged nuclei which, as explained in Chapter 6, may be associated with the semicells becoming slightly separated at the isthmus, but especially by changes in the appearance of the nucleoli. The first perceptible changes in the latter were that it became split into numerous irregular particles which, as the nucleus was clearly entering mitosis, appeared as distinct granules after forming at first, what he described as 'cloudy groups'. Eventually these become rounded, then enlarge by dissolution and fusion and finally congregate into one or more clusters which, he states, are grape-like. In *Micrasterias* these changes occur during the prophase of mitosis, but the so-called 'grape-stage' ends suddenly and strikingly, with a rapid mobilization of the granules. Before this, however, the nucleus is said to assume an opalescent shimmer, which Waris believes may depend on the dissipation of the dissolved nucleolar substance (NS) and on the orientation of nucleoplasm into the spindle. The mobilization of nucleolar particles is the most striking aspect of mitosis in *Micrasterias*, and according to Waris seems to mark the onset of spindle formation, and hence the beginning of metaphase. This type of nucleolar behaviour would seem to be similar to that reported by van Wisselingh (1911) for *Closterium ehrenbergii* (see however, p. 169 below). Waris reports, also in respect to *Micrasterias radiata*, which has a single nucleolus, that this gradually disappears, though on occasions a movement may be observed beforehand.

In a detailed study of the nucleoli in 23 clones from 19 species of 13 genera of desmids, King (1959) has been able to distinguish 4 types of nucleolar arrangement:

i   a single spherical nucleolus
ii   two or more spherical nucleoli
iii a single complex nucleolar mass
iv   a large number of small, spherical nucleoli associated together to form one or more long, bead-like strings.

Apparently, the most common type is the single solitary nucleolus, since this was found to occur in all the genera King investigated including 3 saccoderms, *Mesotaenium*, *Cylindrocystis* and *Netrium* (see, however p. 171). The diameter of the nucleoli varied roughly in proportion to the size of the cells, the smallest being found in *Spondylosium pulchellum* (c. 1 μm diam) the largest in *Netrium digitus* (11 μm).

The Single Spherical Nucleolus

That the single, simple nucleolus is by far the most frequent type is confirmed by Brandham (1964) who found it to occur in 14 out of 18 species of desmids studied. Although most commonly homogeneous when stained with aceto-carmine, some internal differentiation is visible in a few species. Thus in *Closterium siliqua*, *C. acerosum* and *Pleurotaenium trabecula* there is an indistinct and irregular zonation into dense regions embedded in a less dense matrix. In *Cosmarium tetraophthalmum* there is a more regular zonation, which Brandham compares with the NO track described by Godward (1950b) in *Spirogyra*.

In this, and later studies of nuclear division in *Spirogyra* summarised by Godward (1966), it has been shown that the nucleoli are organised by specific chromosomes, the nucleolar organising, or NO regions, invested by a sheath of substance, organiser tracks, having a different staining capacity from the rest of the nucleolar material. They are readily visible within the nucleolus in resting nuclei.

According to King, in the majority of desmids possessing a solitary nucleolus there is no evidence of NO chromosomes, the exceptions being certain clones of *Closterium acerosum*, *Micrasterias truncata* and *Netrium digitus*. In these he reports that fine, thread-like structures are visible within the early prophase nucleolus. Brandham (1964), on the other hand, has shown, from the examination of 19 species of desmids, that there is a very good correlation between the number of NO chromosomes and the number of nucleoli seen at telophase. Only in *Staurastrum tohopekaligense* and *Spondylosium secedens*, which show single nucleoli at telophase, were no NO chromosomes observeable. Also in *Xanthidium antilopaeum* and a clone of *Netrium digitus*, although single NO chromosomes were observed, the number of nucleoli present at telophase was uncertain.

Brandham comments that there would appear to be no correlation between different species within a genus and the number of NO chromosomes, though he has found that species of *Closterium* tend to have more than those of other genera (see also p. 162). He also stresses that one feature possessed by the NO chromosomes of all desmids so far studied, is that throughout prophase they are much more condensed than the other chromosomes and so stain more darkly and have sharper outlines. His study has also shown that the NO chromosomes justify being termed as such, since stages in the condensation of NS on the NO chromosomes have been observed at late anaphase and early telophase. Indeed the most detailed accounts to date, of nucleolar organisation and behaviour during mitosis would seem to be those of Brandham (1964). He has, for example, provided a detailed account of these events in *Closterium siliqua*, a desmid which falls into King's category 1. Its simple spherical nucleolus is 4 μm in diameter (Fig. 67), the nucleus is granular and contains many small chromocentres of which a few, larger than the next, are often seen associated with the nucleolus (Fig. 67: 1). At the onset of prophase the chromosomes become stainable; also the nucleolus becomes less homogeneous and appears as partly fused droplets embedded in a less darkly staining matrix. Eventually thread-like portions of the NO chromosomes can be distinguished. Brandham reports that as the NO threads contract during prophase they become straighter and more easily distinguishable as individual chromosomes. Their morphologies differ, one 7 μm.l* having a long, thread-like region connecting two terminal dot-like regions; the other 4 μm.l is thickened throughout its length except for a 1 μm.l thread-like region near one end. Later the NO chromosomes resemble each other more closely, but before they become too greatly condensed they can often

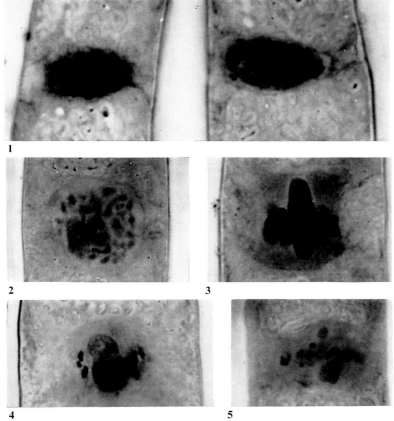

Fig. 67. The nuclear cytology of *Closterium siliqua* West and West—early mitosis.
1 Interphase nuclei in two cells, showing the simple granular nucleus and the single spherical nucleolus. Darkly staining chromocentres are visible within the nucleus.
2 A mid-prophase nucleus, slightly squashed to show the two NO chromosomes. The lower one has 6 darkly staining regions. 3 Premetaphase showing the very considerable elongation of the NS. 4 Premetaphase with elongated NS showing an early stage of its cleavage into two droplets. 5 Premetaphase with the NS now divided into two droplets. (Photomicrographs by Dr P. E. Brandham.)

*μm.l = micrometres in length.

be seen to be composed of several alternately stained and relatively un-
stained regions (Fig. 67: 2). The NS also becomes homogeneous.

The NO chromosomes initially have one end in contact with the margin of the nucleolus and the other end embedded within it. As their contraction progresses it is the inner or free end which moves through the
nucleolar material, the outer end remaining attached and stationary.

The most striking feature of mitosis, at least in desmids of this type, is
that there is no diminution in the quantity of NS, a phenomenon which
occurs in most other organisms. Thus at premetaphase when the spindle
begins to differentiate and the chromosomes move towards the equator,
the NS begins to elongate towards the poles (Fig. 67: 3, 4). Sometimes
in *C. siliqua*, and more frequently in other species, the NS elongates in
one direction only. Brandham records that for *C. siliqua* the maximum
two-directional elongation is 12 μm (Fig. 67: 3). As the premetaphase
events progress towards metaphase itself, the elongated NS begins to con-
strict (Fig. 67: 4) and eventually divides into two droplets which migrate
towards the poles of the spindle (Figs 67: 5 and 68: 1). The cleavage of NS
is usually unequal, and Brandham records that the NO chromosomes
remain attached to the NS throughout (Fig. 68: 1). By late metaphase the
NS has completely dispersed.

The spindle of many desmids is also quite unusual, and as can be seen
in Fig. 68: 2, 3, in *C. siliqua* it is cylindrical and truncate (13 μm.l ×
10 μm.br*). Parallel disjunction of chromatids has been observed at
anaphase, though the NO chromosomes are frequently sticky and form
temporary bridges due to the delayed separation of the chromatids.
Brandham illustrates more complex structures, and these he interprets as a
trailing chromatid of one NO chromosome attached to a sticky bridge,
caused by the other, which thus forms a Y-shaped structure (Fig. 68: 4). By
mid-anaphase, bridges are no longer apparent though the chromatids of
the 2 NO chromosomes are often seen trailing behind (Fig. 68: 5).

The NS begins to condense by late anaphase and reappears as droplets
scattered about the anaphase plates and not within them, as is most
common, and, as might be expected, is associated with the 2 NO chromatids (Fig. 68: 5). The chromosomes quickly lose their identity at telophase,
though the NO chromosomes are distinguishable for slightly longer.
Either one or two nucleoli (Fig. 68: 6) are formed at telophase, a fact
which seems to depend on how close the 2 NO chromosomes are to
each other. It is while still in the telophase condition that each daughter
nucleus carries out its remarkable migration round the base of the chloroplast and along the convex side of the cell (cf. p. 107, and Fig. 68: 6), to
their positions halfway along the length of the two dividing chloroplasts.

*μm.br = micrometres in breadth.

By the end of the migration, there is always but a single nucleolus within each nucleus, so that if two are initially formed, clearly they must fuse during this event.

Brandham's (1964) investigation of *Gonatozygon kinahini* has revealed

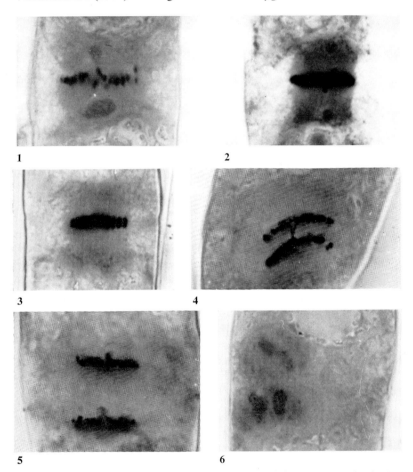

Fig. 68. The nuclear cytology of *Closterium siliqua* West and West—late mitosis. 1 Metaphase with the droplets of the NS at the poles of the spindle. The upper droplet is still attached to an NO chromosome. 2 Metaphase, showing the typically truncate spindle and the still diminishing droplets of NS. Both of the NO chromosomes are orientated at right angles to the metaphase plate. 3 Late metaphase by which time the NS has completely disappeared. The pairs of chromatids are clearly visible. 4 Anaphase showing a trailing NO chromatid; also a bridge caused by the delayed division of the other NO chromosome. 5 Late anaphase with NS recondensing around the chromatid plates. Most of it is associated with the NO chromatids. 6 An early stage in nuclear migration when each of the migrating nuclei contains 2 nucleoli. (Photomicrographs by Dr P. E. Brandham.)

that this species has a single irregularly shaped interphase nucleus with a number of small chromocentres (more were seen in *G. monotaenium* by King). The single nucleolus is sub-spherical and about 5 μm in diameter. At the onset of prophase, the chromocentres enlarge and a single NO chromosome becomes visible, embedded in the NS. The chromosomes approach the equator of the developing spindle at premetaphase and the NS partially dissolves and diffuses throughout the nucleus. The rest of the NS elongates towards the poles of the spindle, pulling the NO chromosome with it and causing it to become, as in many desmids, orientated at right angles to the plate. The NS does not, however, divide, but dissolves away in this position and hence is unusual (cf. p. 172). Brandham observed trailing chromatids of the NO chromosome at anaphase, and comments that as anaphase progress into telophase, the daughter nuclei are reconstituted. However, they do not resemble typical interphase nuclei until their migration is well advanced.

Two or more Spherical Nucleoli

King's second category of two or more spherical nucleoli per nucleus was reported by Brandham (1964) in only one, clearly unusual clone, of *Netrium digitus*. Most frequently he found 2 or 3 per nucleus but in one cell he found 17. He comments that King's second type is rare and found only in diploids of *Cosmarium botrytis* and in *Micrasterias americana*. In the latter from 1 to 3 nucleoli are visible though occasionally many can be seen. In a large strain of *Closterium acerosum*, Brandham observed 3 nucleoli and so it was possibly a polyploid (cf. p. 171), while investigations of 3 different clones of *Netrium digitus* showed, in contrast to King, that the nucleoli of each was complex and that they differed from one another. Thus he describes one as a loose aggregation of bodies; the second was composed of globules, which may have been part of a single coiled body. (cf. *Cl. moniliferum*, p. 166); third contained nucleolar membranes.

The Complex Nucleolar Mass

King's third category has been described so far only in *Closterium moniliferum* though it could be more widespread. For example, two clones of *Netrium digitus* are said by Brandham (1964) to resemble this type, though he has not been able to demonstrate unequivocally the presence of a single coiled thread. King's careful scrutiny of a large number of *C. moniliferum* in interphase and early prophase revealed that the nucleolus appears to be an elongated body of varying diameter and coiled tightly around itself, a feature which he states gives each a lobed appearance. Brandham (1964) found that in certain interphase nucleoli structures of an apparently membranous nature can be observed, which he states 'have a sinuous distribution and indicate a complex internal structure . . .' (Fig. 69: 1, 2).

King (1959, 1960) is of the opinion that these all form part of the process of nucleolar dissolution at prophase or its reformation at telophase. Brandham (1964), however, also records their occurrence in interphase nuclei

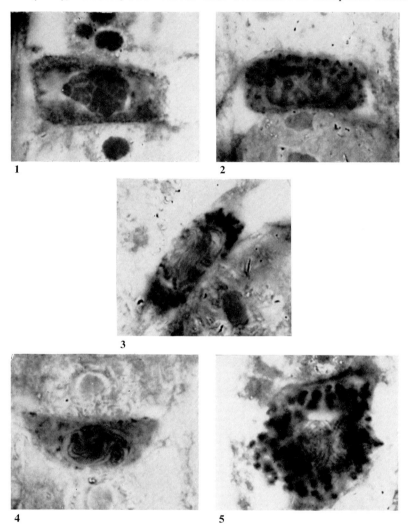

Fig. 69. The nuclear cytology of *Closterium moniliferum* Ehr.—'nuclear membranes'.
1 Interphase nucleus showing many small chromocentres and the compound nucleolus. 2 Interphase nucleus showing details of its folded membranes. 3 An early prophase nucleus with the nucleus entirely in the form of 'membranes'. 4 Later early prophase with the nucleolar membranes becoming less clear. Droplets of NS are appearing and some of the chromosomes are divided into chromatids. 5 Another early prophase nucleus in which the chromocentres are beginning to enlarge. Nucleolar membranes are associated with large droplets of NS.

(see also Godward 1953, 1956, Newnham 1962) and so disagrees with King's interpretation. In support of this view Brandham fixed *C. moniliferum* at intervals during the 'light period', a time when mitotic stages are rarely found. Slide preparations were made of each fixation and the percentage of cells showing 'nucleolar membranes' determined. Table 8 shows his results.

Table 8. Percentage of cells of *Closterium moniliferum* showing 'nucleolar membranes' at different times during the 'light period'.

Time of fixation (hours)	Percentage showing 'membranes'
10.00	36.3 ± 2.9
11.00	33.6 ± 3.7
12.00	33.7 ± 3.4
13.40	32.5 ± 3.0

Clearly the percentage of cells exhibiting 'nucleolar membranes' remained constant over the period when the cells were fixed. Although it could be argued that they are part of an extended prophase, Brandham states that this seems unlikely in view of Waris' (1950) observations, and those of Brandham himself (Brandham and Godward 1965b), on the time taken to undergo mitosis. Even if the prophase took 4 hours to complete (the longest time noted by Brandham, and that for *Micrasterias thomasiana*) it is not unreasonable that some of the cells in prophase at 10.00 hours should show later mitotic stages before 13.40 hours. Since no such stages were observed it was therefore concluded that the 'membranes' are most probably associated with the interphase condition.

Brandham observed 'membranes' in nuclei which were clearly in prophase because of the appearance of the chromosomes (Fig. 69: 3). Thus he decided to determine whether these prophase 'membranes' were merely relics of those observed at interphase, or whether they were also in fact produced at prophase in nuclei which showed none at interphase. If the former was the correct interpretation then the percentage of prophase nuclei containing nucleolar 'membranes' should approximate that of interphase nuclei containing them. If they were produced at prophase then their percentage occurrence should be greater. In a slide containing mitotic stages the percentage of interphase nucleolar membranes was 28 % while 58.8 % of the prophases examined contained membranes. Hence, Brandham concluded that membranes are produced during prophase in addition to those present in interphase nuclei. Although the exact nature of nucleolar membranes is still not known, Godward (1953), who found them in *Spirogyra* spp., has suggested that they are phase boundaries between reactants. Brandham (personal communication) has commented that they may even be in some way connected with viral infections of the cells in which they occur.

The 'membranes' are always associated with dense masses of NS in which large chromocentres belonging to the NO chromosomes can be seen to be embedded (Fig. 69: 4). As prophase proceeds the limits of the nucleolus become increasingly vague though a mass of membrane-like structures remain (Fig. 69: 5). Gradually, however, these disappear, but as they do so, droplets of NS appear, embedded in a less dense matrix, also of NS (Fig. 70: 1).

By mid-prophase 3 NO chromosomes become visible and these are morphologically distinguishable one from the other and they are usually closely associated with the denser droplets of NS. At premetaphase the NS migrates either to one spindle pole, or, as happens most frequently, it elongates towards both poles (Fig. 70: 2). Unlike the NS in *C. siliqua*, however, the elongated NS does not split into two, but seems to dissolve while still elongated, and by late metaphase it has quite disappeared.

At anaphase, parallel disjunction of the chromatids has been observed and during this stage the spindle changes its shape from cylindrical to barrel-shaped (Fig. 70: 3) or even sub-spherical.

Telophase proceeds in a similar manner to that observed in *C. siliqua* except that in most cases the NS condenses inside the chromatid plates. Three nucleoli are formed in which portions of NO can be seen. The nucleoli fuse on the completion of the nuclear migration, sometimes forming 'membranes' similar to those of prophase or interphase nuclei.

Nucleoli as Bead-like Strings

Nucleoli composed of very many small bodies in linear association, King's fourth category, are according to him, confined to the genus *Micrasterias* (see, however, Kasprik 1975 and p. 171). For example, in *M. thomasiana*, King counted over 100 nucleoli in a single nucleus, and even in the interphase nuclei of living cells these, he states, could be seen clearly to be associated in long bead-like strings. *M. rotata*, however, has only 30 or so nucleoli and their string-like association did not become evident until prophase was fairly well advanced. During prophase in both species the strings, of which there are 2 per nucleus, shorten, in consequence of which the nucleoli become drawn closely together and eventually become so closely adpressed that they appear to coalesce into a continuous mass. There is a gradual dissolution of the masses until all that remains are two long chromosomes which King presumes to be NO chromosomes. At a late stage in the dissolution he reports that one each of

M

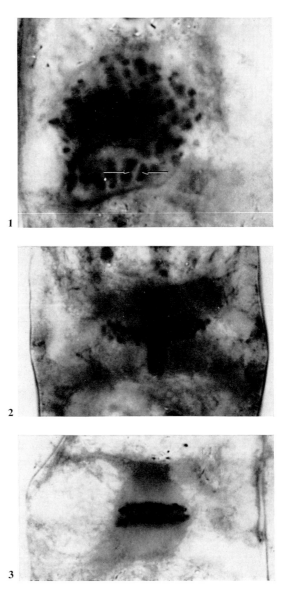

Fig. 70. Nuclear cytology of *Closterium moniliferum* Ehr.—later stages in mitosis.
1 Prophase in which nucleolar membranes are becoming less clear and droplets of
NS are beginning to appear. Arrows point to chromosomes which have divided into
chromatids. 2 Early metaphase showing cylindrical spindle and elongation of the NS.
3 Anaphase in which the spindle has become barrel-shaped. Spindle fibres visible.
(Photomicrographs by Dr P. E. Brandham.)

these chromosomes can be seen running throughout the length of each body.

King points out that the nucleolar behaviour which he observed during mitosis in both *M. thomasiana* and *M. rotata* is by no means in agreement with the descriptions given by Waris (1950) for these two species (see p. 159 above). This is especially remarkable, since identical clones were used in both studies. It may however be significant in this connection that Brandham (1964) found the nucleolus of *M. americana* to be variable even within a single clone. Thus it may consist either of one large body 6 μm in diameter; two or three smaller bodies partly fused together; or a mass of very small bodies which resemble those present in *Closterium ehrenbergii*, though they are more widely scattered.

Nucleolar Variability

Details of the structure and mitotic behaviour of King's fourth type of nucleolus are reported by Brandham (1964) in *Closterium ehrenbergii* and one clone of *Netrium digitus*, a possible polyploid. In *C. ehrenbergii* the stained interphase nucleus is lens-shaped and 60–65 μm in diameter and 30 μm thick. When stained it appears as a fine reticulum in which numerous chromocentres are scattered. The nucleolus is a mass of droplets each 4–8 μm in diameter loosely aggregated to form a complex lens-shaped structure about 25 μm.l by 18 μm thick (Fig. 71: 1). During prophase, chains of chromocentres appear and gradually these shorten and condense. The nucleolus becomes more irregular and droplets become smaller and more numerous (Fig. 71: 2), and Brandham has found chains of droplets similar to those found by King in *Micrasterias rotata*. The NS in *C. ehrenbergii* is in the form of widely scattered droplets of 1–2 μm diameter. At early metaphase part of the NS can be seen at the poles of the spindle, the rest migrating towards them. By late metaphase all the NS reaches the poles but has diminished in amount, and, in contrast to other *Closterium* species, the NS at the spindle poles persists throughout anaphase, though its volume sometimes seems to diminish. Temporary bridges are seen between chromatids in early anaphase and the longer chromosomes (NO?) trail. Brandham emphasises that the major difference between the anaphase of this and all other desmids described so far is that the chromatids are U, V, or J shaped (Fig. 71: 3); an indication that unlike other desmids (see p. 172), possibly localised centromeres are present in *C. ehrenbergii*, though this has not yet been substantiated.

Different clones of the same species may have nucleoli which differ significantly from one another. Thus in one clone of *Netrium digitus* Brandham (1964) found it to consist of a loose aggregation of bodies

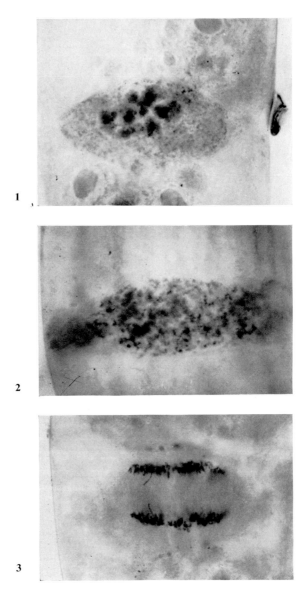

Fig. 71. The nuclear cytology of *Closterium ehrenbergii* Menegh. 1 An interphase nucleus showing the complex, compound nucleolus. 2 Early prophase with chromo-centres and showing the nucleolus becoming scattered throughout the nucleus as; 3 mid-anaphase showing droplets of NS still present at poles of spindle. The chromatids of one long chromosome are trailing behind the plates. The other chromatids can be seen to be U-, V-, or J-shaped. (Photomicrographs by Dr P. E. Brandham.)

similar to the nucleolus of *Closterium ehrenbergii* (King's Type iv), in another it resembles that of *C. moniliferum*, though Brandham is uncertain as to whether its globules were part of a single coiled body (King's Type iii). In a third clone there may be a single, simple nucleolus (King's Type i), or a few simple bodies (King's Type ii). As Brandham comments, 'it is, to say the least, remarkable that a range of nucleolar structure embracing the simplest to the most complex' should be found within a single species (cf. Kasprik 1975 and p. 172).

That nucleolar structure varies within a given genus is supported by a study of the nuclear cytology of the genus *Micrasterias* by Kasprik (1975). After examining some 54 strains of 23 species or varieties of *Micrasterias* he reports that the size, form and number of particles constituting the nucleoli varies from species to species, though in general there is a direct relationship between cell and nuclear size, and nucleolar mass. Relatively small species, such as *M. truncata*, can, however, have a markedly dispersed nucleolus. Kasprik recognises within the genus two major types of nucleoli. The first is the single, more or less rounded, closed nucleolus which occasionally exhibits a furrowed surface (*M. papillifera*) and vacuolar inclusions as found in *M. furcata* and *M. radiata;* also in some clones of *M. conferta*, *M. muricata*, and in *M. papillifera*, *M. furcata* var. *gracillima*, *M. swainii* and *M. tetraptera* var. *taylorii* (King's Type i?).

Kasprik's second category of nucleoli in the genus are the complex nucleoli which he separates into two sub-categories. One is the elongated nucleolus which he describes as being coiled, so that the units comprising them are usually difficult to count (King's Type iv?). Some of these, such as in *M. rotata* have a relatively smooth outline so that they appear thread-or worm-like. Others, characterised by *M. thomasiana*, consist of aggregates of extremely linearly structured particles (cf. King 1959 and p. 167). The other sub-category is recognised as having a central nucleolar mass with what Kasprik describes as a brain-like, furrowed surface and gives as examples *M. fimbriata* var. *spinosa*, *M. papillifera* and *M. furcata*. Although Kasprik does not make the point, since he does not describe the behaviour of any nucleoli during mitosis, it must be presumed that this sub-category of complex nucleoli corresponds to Kings' Type iii.

Kasprik was able to find examples of King's '2 or more independent nucleoli' category only under precisely defined culture conditions. It may be significant that in cultures, as pointed out by Drawert and Mix (1961), cell age may have quite a profound effect of nucleolar organisation. Kasprik (1975) also comments that he was unable to resolve with any clarity in *Micrasterias*, the relationship between the nucleolus and the NO chromosomes. In fact he poses the question as to whether the nucleolar organisers of the Zygnemaphyceae represent a unique structure for, as he states, typical secondary constrictions could not be found in *Micrasterias*, even though some satellite-like segments may appear in some stages of mitosis. According to Godward (1966) conjugate nucleoli consist of a basic matrix, the nucleolar substance, which disperses during mitosis and then reassembles, plus the organiser tracks. This distinction could not, however, be discerned according to Kasprik in *Micrasterias* unless, as he states, one recognises the lightly staining zones which also surround the nucleolar segments as an additional component of the nucleolus. As he points out, these could be fixation artefacts.

This fact has been particularly emphasised by Drawert and Kalden (1967) in a study of the nucleolar changes in cultures of *M. rotata*. At the end of telophase they could usually see 3 primary nucleoli per nucleus. Their 'primary form' was spherical, but during the growth of newly forming semicells it was found to change into a thread- or worm-like 'secondary form'. This, they state, is the condition typical of physiologically active cells, growing in more or less optimal culture conditions. If growth conditions become less favourable, the worm-like interphase nucleoli contract and become compacted into forms that they compare to pieces of coke. Fragmentation of these nucleoli may occur at this stage, hence leading to the formation of smaller nucleolar bodies. Should culture conditions remain unfavourable these may reunite to form larger bodies. The fusion of nucleoli and smaller bodies eventually results in the production of one homogeneous, spherical, secondary nucleolus and appears to be indicative of cell degeneration which is irreversible. If, however, this fusion stage has not been reached, the addition of fresh culture medium can cause the worm-like, secondary nucleolar form to reappear.

Kasprik suggests that the ultimate character of nucleolar form is probably physiologically conditioned and does not constitute any evolutionary progression from simple to complex. Clearly there is much more to be learnt about the complexities of the desmid nucleolus.

2.   THE CHROMOSOMES

In the early investigations of desmid chromosomes interest was focussed largely on chromosome numbers rather than on their morphology (van Wisselingh 1911, 1912). However, stimulated by the researches of Godward (1950a, b, 1953, 1954a, b) on *Spirogyra*, which revealed many unusual chromosomal features, including the presence of NO regions already alluded to and non-localised centromeric chromosomes, attention was directed towards desmids, especially by Godward's students (King 1960, Brandham 1965a).

King developed a squash method for making preparations of metaphase plates enabling them to be seen in polar view, since, most frequently, the dividing cell presents a lateral view to the observer. He examined, with this technique, the chromosome numbers and morphology of both sacco-derm and placoderm desmids of 14 different species. In the case of *Netrium digitus* he studied 2 and in *Cosmarium botrytis* 3 different clones. The lengths of the chromosomes at metaphase ranged in length from less than 0.5 to 4.5 μm, though most were less than 1.5 μm. In general, chromo-some size was directly related to cell size; it was also found to be inversely proportional to the number of chromosomes present per nucleus. For example, in the 3 clones of *C. botrytis*, which had distinct chromosome numbers, their lengths were as follows:

$$\text{Clone A } n = 18 \quad \text{length 2.0–3.5 μm}$$

$$\text{B } n = 26 \qquad\qquad \text{1.0–2.5 μm}$$

$$\text{C } n = 94 \qquad\qquad \text{less than 1.0 μm}$$

Because desmid chromosomes are on the whole so small, and also invested in a darkly staining matrix, King was able to distinguish little of significance about their morphology and he described them 'as either dots or rods, clearly divided *throughout their length* (his italics) into chromatids'. He found, however, that in some species the chromosomes during mid-prophase can be seen to possess a number of stainable regions (usually not more than 5) intercalated with less stainable zones, which give them a banded appearance. However, as metaphase approaches and the chromo-somes contract, the stainable bands merge, so that at metaphase the banded appearance can no longer be seen. In *Micrasterias rotata* the NO chromo-somes (see also p. 167) which are longer than the others (4.5 μm), remain banded throughout metaphase and anaphase.

The most recent study of desmid chromosomes is that of Kasprik (1975) in which he examined 23 species of *Micrasterias*. In this genus he found various karyotypes differing not only in chromosome number but also morphology. He developed a new and apparently successful micro-dissection technique whereby metaphase plates were extracted undamaged so that exact chromosome counts could be made. As a result of his survey he was able to define 4 different groups on the basis of their chromosomes, as follows:

Group 1   Species with small chromosomes, mostly 1 μm, which because of the stickiness of their matrix tend to aggregate: these include *M. crux–melitensis, M. pinnatafida, M. furcata* and *M. truncata*. Species of the *M. americana–M. laticeps* group show less tendency to aggregate.

Group 2   Well-separated chromosomes with relatively distinct contours which stain intensively and homogeneously. This type is seen at its best in *M. papillifera*, *M. rotata*, *M. swainii* and *M. tetraptera* var. *taylorii*. However, some species do show a slight tendency to aggregate. The chromosomes are about 1 μm.l, though the NO chromosomes may be significantly longer.

Group 3   Still relatively short chromosomes up to 2.5 μm.l but which are rather thick, and so appear compact. Their structure, however, is less compact than Group 2 chromosomes, and even when fully contracted, they often contain regions which are less stained than the rest; their outlines are somewhat hazy. The following species have chromosomes of this type; *M. brachyptera*, *M. denticulata* var. *angulosa*, *M. fimbriata*. *M. muricata* stands somewhat apart in having larger, less clearly delimited, but strongly aggregating, chromosomes.

Group 4   Typified by longer (up to 7.5 μm), compact chromosomes which appear to be joined together. As well as *M. torreyi* perhaps also *M. thomasiana* should be included in this group, even though in its complement shorter ones predominate. *M. radiata* is different in having longer, less clearly limited but strongly aggregating chromosomes.

Giant chromosomes have been found by Gerrath (1966) in *Triploceras verticillatum* ($n = 15$). In studies of mitosis of this desmid he reports length of up to 20 μm in late prophase preparations. He also found the *T. gracile* has chromosomes much larger than those normally found in desmids. As stated earlier, long chromatids which are V- or J-shaped, and hence possibly possessing localised centromeres, have been found at anaphase in *Closterium ehrenbergii* (Brandham 1964; see p. 171 and Fig. 71). The late prophase chromosomes in this species are said by Brandham to be 2–5 μm.l. It must be noted that Brandham and Godward (1965) in their study of the ultrastructure of the nucleus of *Closterium ehrenbergii* could find no evidence for the presence on its chromosomes of a localized point of attachment of the spindle fibres.

Except in the case of the saccoderm *Mesotaenium caldariorum* and *C. ehrenbergii*, in which there is a normal disjunction of chromatids at anaphase, with one point of each chromatid leading the way to the pole, all the desmids which King and Brandham examined undergo parallel disjunction (King 1953a). This type of separation was described by both Geitler (1935) and Godward (1950a, 1954) in *Spirogyra* and has been found to occur in a few higher plants and some animals. X-ray produced chromosomal fragments of such chromosomes have shown to move independently to the poles (Hughes-Schroeder and Ris 1941) thereby confirming the view that parallel disjunction is a consequence of a non-localised centromeric organisation. Indeed the question of the centromeric organisation of the desmid chromosome remains unsolved. There is some evidence for the existence of 'diffuse' or polycentric chromosomes in a number of desmids (King 1953a, 1960, Brandham 1965, Brandham and Godward 1965), though Ling and Tyler (1976) and Ueda (1972) have observed median, centromeric constrictions in others.

Godward (1966) has compiled a table (her Table 1.91) of chromosome numbers in desmids, many from counts made in studies by her students (King 1960, Nizam 1960, Brandham 1964) and others from earlier investigations (Acton 1916, Karsten 1918, Pothoff 1927, Waris 1950, Maguitt 1925, Kopetzky-Rechtperg 1932, van Wisselingh 1912). The numbers shown range from 14 in *Staurastrum gracile* to one clone of *Netrium digitus* with cells differing in number from between 172–182 to 592, the highest chromosome number ever recorded for one nucleus (King 1953b).

In common with other members of the Zygnemaphyceae, desmids are normally haploid in the vegetative condition, only the zygospore being diploid. Meiosis occurs at zygospore germination restoring the haploid number (Brandham and Godward 1963, see also p. 182). As King (1960) points out, chromosome numbers in desmids have, in the main, been obtained from cells which have been in clonal culture for several weeks and irregularities of mitosis consequent on cultural conditions would seem to be responsible for the development of chromosomal races. For example, Brandham (1964) examined 5 clones of *Cosmarium botrytis* and found numbers characteristic of an aneuploid series—20, 21, 24 and 26. Similarly in *Netrium digitus*, in addition to the very high numbers mentioned above, King (1960) also recorded counts of 30 and 32 in different clones. Such aneuploid variation is a fairly common feature of plants possessing non-localized centromeric chromosomes e.g. *Luzula* (Nordenskiöld 1957) and *Carex* (Heilbron 1939, Tanaka 1939, Wahl 1940). King (1960) suggests that the intraspecific differences in chromosome numbers has been brought about largely by transverse divisions, or fusions of chromosomes. The possession of non-localized centromeres would allow fragmentation to occur without loss of the resulting fragments. Culture conditions may have increased the incidence of fragmentation (and fusion) of chromosomes and similar events may occur in natural desmid populations, though less frequently.

3. POLYPLOIDY

Changes in the chromosome complements of desmid cells have been induced in the course of the now classic investigations of Waris and Kallio on desmid morphogenesis summarized in their paper of 1964 (see p. 187).

Hence, in one study designed to explore nuclear–cytoplasmic relationships (Waris and Kallio 1957), they artificially produced stable polyploid forms of numerous *Micrasterias* species by various means. In most cases they arose from the fusion of two adjoining nuclei produced from 'complex cells' (p. 121) after the disturbance of cytokinesis by cold shock or heat treatment, or the use of dinitrophenol and centrifugation (see Fig. 53). Tetraploid nuclei have similarly been produced from diploid nuclei and triploids from 3 haploid nuclei. The viability of tetraploids was low, and no permanent, constant clones resulted. Only rarely in *Micrasterias* did diploid nuclei result from a disturbance in mitosis after the chromosomes have doubled, but this did occur following plasmolysis and cold treatment (Kallio 1951); also when haploid cells in metaphase were transferred to Waris' (1953) M.S. solution containing 10^{-3} mole sodium 2,4-dinitrophenolate per litre for 4 hours.

When cells of *M. rotata* var. *evoluta* were centrifuged at metaphase the nucleus, on occasions, became divided into more than two daughter nuclei, a phenomenon termed 'nuclear splitting' (Waris 1958). By this means, nuclei were obtained, one which was normal and two small. In one special case, following centrifugation, one of the two anaphase groups had become severed by a new septum giving rise to two different nuclei. As a consequence, one of the daughter cells received a normal and a deficient nucleus; the other only a deficient one. After fusion of the two unequal nuclei in the former instance, a hyperhaploid clone resulted, while the cell with the deficient nucleus survived for only a month and did not divide. The hyperhaploid clone had some of the features common to diploid clones (see below) in that there was some lobe duplication. They were, however, not as stable as diploids and after several years in culture Waris and Kallio report that they have reverted to the normal haploid form.

Diploid cells are always larger than haploids in all *Micrasterias* species in which this condition has been induced. There are, however, two different types of diploid cells; one with the characteristic biradiate facies but significantly larger than the normal haploids which they resemble; the other possesses additional side lobes, or is tri- or even quadriradiate. Thus, for example, in diploids of *M. thomasiana* all the side lobes usually form sublobes at two different levels; also a quadriradiate facies has been produced in *M. thomasiana* whose wings, interestingly, are similar in form and size to those of normal haploid cells.

In all species of *Micrasterias* studied by Waris and Kallio, the morphological variation exhibited by diploids was found to be greater than in haploids. Also the diploids were much more sensitive to environmental conditions than haploids. Thus the temperature maximum for clone growth is lower for diploids, and a temperature of 25°C was found to kill diploid *M. thomasiana* cells, but not haploids. When grown in the same culture vessels as haploids, diploids were always found to divide at a slower rate (Kallio 1951, 1961, see also Brandham and Godward 1965 and p. 121).

Brandham (1965) made a special study of polyploidy in *Closterium siliqua, Cosmarium botrytis, Staurastrum denticulatum* and *S. dilatatum.* After establishing cultures of these desmids he found in the case of *C. siliqua* that a few cells were considerably larger than those comprising the majority of the population—260 μm.l compared with most at 162 μm.l. Cytological preparations of both types of cells revealed that the larger cells possessed nuclei much larger than those of the smaller cells which had chromosome numbers of $n = 86$. Also in the latter nuclei there were 2 large NO chromosomes, and although it was not possible to count accurately the chromosomes in the larger cell nuclei, 4 NO chromosomes were seen which is sufficient evidence to indicate that these were almost certainly polyploids.

Brandham was able to reconstruct how such diploids arose from haploid cells. When large cells were found amongst the normal small cells, he observed a number of monstrous bilobed cells resembling two normal cells joined end to end by a narrow isthmus. Cytological examination showed that these contained either 2 haploid nuclei lying close together in one lobe, or a single diploid nucleus. These cells he presumed to be formed by nuclear displacement, before or during mitosis, with the result that two daughter nuclei were on the same side of the newly formed septum (see also Kallio 1951, Waris and Kallio 1957, 1964).

Following UV radiation in which 90% of the population of *Cosmarium botrytis* in clonal culture were killed, some abnormally large and some abnormally small cells were found among the survivors. One of these large cells was successfully raised in culture and found to be diploid. Interestingly, its growth rate was significantly slower than haploid clones (cf. 121). This culture also began to produce a number of triradiate cells, and after a few months the percentage of these cells amount to 97%. The diploid cells were on average 77 μm.l compared to the haploids which were 64.4 μm.l; also the nuclei of the former were correspondingly larger. The chromosome number of the haploids was $n = 20$ with one nucleolus, while the diploids which, because of their sticky nature were difficult to count ($2n = $ ca 42), were frequently seen to possess 2 nucleoli.

Presumed diploids of *Staurastrum denticulatum* arose spontaneously amongst germinating zygospores produced by conjugation within the original homothallic haploid clone. It is again of interest to note that as cultures derived from isolates of these diploids developed, they became, in a few weeks, almost exclusively quadriradiate (cf. p. 120). The triradiate haploids measured 26.1 μm.l while the quadriradiate diploids were 40.2 μm.l; also there were some observed differences in morphology.

From a non-sexual population of *S. dilatatum*, which contained both 3, and 4-radiate and 3/4 Janus cells (23.75 μm.l on average), larger cells (34.5 μm.l) were found in the clone. When isolated, the resulting population consisted entirely of large, mostly 5-radiate cells, with a few 6-radiate. Brandham (1965c) inferred that the larger cells were diploids, or at least hyperhaploids. A very large cell, 49.5 μm.l, was seen in one of the diploid clones and this possessed two whorls of 8–10 processes on each semicell. Brandham presumed that this cell represented an even higher degree of polyploidy and suggested that it might possibly be a tetraploid.

Kasprik (1975) reports an interesting case of an aneuploid series correlated with morphological differences in *Micrasterias thomasiana*. In cultures of this desmid, clearly distinguishable variants, alongside normal forms were found to be hyperhaploid. The basic chromosome number of this desmid is $n = 39$, but one variant, in which there was an overlapping of the lateral processes, contained $n = 40$, while a second of even more irregular form was found to have $n = 46$. Two further variants which were triradiate had chromosome complements $n = 70$ and $n = $ ca 75. Some of the cells of this latter clone were inclined to develop more or less typical *M. thomasiana* morphologies, except that they were significantly larger. Frequently biradiate cells were found to arise from triradiate forms, but these exhibited an overlapping of their semicells in the isthmus region, a condition described also by Kallio (1951). Changes in the degree of radiation in some desmids may result from an increase in the level of ploidy, as concluded by Starr (1958) in his study of a heterothallic strain of *Cosmarium turpinii* which produced not only normal biradiate, but also tri- and quadriradiate cells. Evidence that polyploidy was involved came from cytological observations, in which he found between 30–36 bivalents in normal strains at diakinesis, while the protoplasts from germinating zoospores from crosses between normal and triradiate strains showed univalents and trivalents in addition to the anticipated bivalents. Chromosomes counts in these produced total numbers in the region of 60–72. Counts of large triradiate forms, though not accurate, gave clear indications that these were diploid. 'Hybrid' zygospores which failed to produce any viable offspring following meiosis were probably triploids, or even tetraploids, depending on the nature of the parental strain used in the original crosses.

Starr considers the production of large forms of *C. turpinii* to be a response to an increase in nuclear quantity; the change in shape from biradiate to tri- or quadriradiate, a response to the increase in cell volume, rather than a specific gene-controlled factor capable of being inherited by the progeny and resulting from sexual reproduction. Aspects of this problem will be considered in greater detail in Chapter 8 on morphogenesis.

The contrast to polymorphism resulting from polyploidy is that found by Brandham and Godward (1964) in *C. botrytis*, where the volume of triradiate was no greater than that of biradiate cells, both types being presumed to be haploid. Similar, small triradiate forms were also found by Starr in his study of *C. turpinii*. Brandham and Godward explored the inheritance of haploid triradiate cells in *C. botrytis* (see also Tewes 1969). Although cultures of this desmid were maintained for many years, triradiate cells were never found amongst vegetatively reproducing material, but they did occur quite commonly, as Starr also found, in populations derived from germinating zygospores. In the case of Starr's *C. turpinii*, it arose only from one of the two cells produced per germinating zygospore, whereas in *C. botrytis* it arose from either; and in one or both of the pair of juvenile cells.

In newly isolated cultures of triradiate clones all cells were similar and symmetrically triradiate, but within a month biradiates were found and these increased, because of their faster growth rate, to eventually swamp out the triradiates. An additional very interesting observation was the fact that triradiate cells having lobes of unequal size also appeared, and Brandham and Godward conclude that such 'reduced' cells are intermediates produced by the gradual loss of the triradiate condition. Evidence that there was a gradual change, in contrast to the abrupt alteration in radiation in which Janus forms (2/3) occur as in *Staurastrum chaetoceras* (Reynolds 1940) or *S. sebaldi* var. *ornatum* (Lind and Croasdale 1966), was that both semicells were of the same 'reduced' triradiate type. As also found by Starr (1958) when such reverting clones were isolated, the resulting culture was entirely biradiate, thus indicating a permanent loss of the triradiate character.

To examine whether the triradiate form was inherited through the sexual process, Brandham and Godward showed that the triradiate clones were found to be fully fertile, but that the character was not transmitted sexually. They argued that if triradiation is a gene mutation inherited independently it would occur in half the progeny of a 'triradiate' × 'normal' cross. If, however, it was to be thought of as a gene, linked to that of a mating type, it would occur in all the progeny having 'minus' mating type, except in the case of recombination. Their results showed that neither cases applied, thus ruling out genetic control, since the triradiate character was found to occur no more frequently in the progeny of a 'triradiate' × 'biradiate' cross.

Polyploidy is suspected, though not yet confirmed cytologically, in the genus *Pleurotaenium* by Ling and Tyler (1974). Their observations on conjugation in *P. ehrenbergii*, *P. mamillosum* and *P. coronatum*, and resulting conclusion that these taxa are members of the same biological species, have been summarized in Chapter 6 (p. 152), as has the fact that *P. coronatum* is probably a diploid. Just as Brandham (1965) showed with *Cosmarium botrytis* that conjugation between diploid and haploid cells involves 1 diploid and 2 haploids, so also Ling and Tyler observed that when *P. mamillosum*, or *P. ehrenbergii* are crossed with the much larger (diploid?) cells of *P. coronatum* large numbers of the presumed haploids associate during conjugation with a single *coronatum* cell. Such multiple pairings suggest to them that at least *coronatum* is a 'hyper-haploid', producing a powerful sex attractant.

The conjugation of distinctly heteromorphic cells of *P. subcoronulatum* has been reported by Ramanathan (1964) from Madras, India. The large, cells were 200–260 µm.l, while the short ones with which these paired were as small as 40–50 µm.l. Also individuals were observed in which one semi-cell was fully developed while the adjacent one was short. The smaller cells were always separate individuals, but the long ones frequently occurred, and underwent conjugation with the former, as filaments. Ramanathan, commenting on this morphological anisogamy, believes that a sexual differentiation exists, the long cells corresponding to females, the short to males (cf. Geitler 1958).

CHROMOSOMES AND DESMID TAXONOMY

Clearly the 3 *Pleurotaenium* 'morphs' studied by Ling and Tyler have the ready ability to share a common gene pool, and so they suggest the synonomy of the 3 taxa with *P. ehrenbergii* taking precedence. As they suggest, it seems very likely that many so-called desmid species exist in a number of environmental varieties, each currently bearing specific rank but which, when genetic criteria are adopted in delimiting species, may be found, as in the case of their *Pleurotaenium* 'species', to be untenable as such.

Kasprik suggests when differences in chromosome length and number occur, as King (1960) found in different clones of *Cosmarium botrytis*, that agametoploidy (cf. *Luzula*) must be suspected. He found similar differences in clones of *M. americana* but with only small differences in chromosome size. Thus his lowest counts were $n=88$ and 93; an intermediate number was $n=135$ and these had distinctly smaller chromosomes; the highest count was $n=c.\ 205$ but the chromosomes of this clone appeared to be no smaller than those of the $n=135$ clone. Kasprik is uncertain as to the extent to which fusion, or fragmentation, has led to the observed differences in chromosome size and number. Also the question, 'which is the original chromosome type?' must remain open; though the predominance of small chromosomes, such as found in *Micrasterias* and *Cosmarium*, may well represent the primitive condition. Attempts to reorganise species groups on the basis of chromosome size would seem to be, at least at this stage of our knowledge, unwarranted. For example, the *Micrasterias* species with the largest and most complex cells (*M. denticulata*, *M. rotata* and *M. torreyi*), which might be expected to have some affinity, in fact possess 3 quite distinctive chromosome complements. On the other hand, Kasprik emphasises that in the evolution of karyotypes, simple quantitative changes do not appear to lead to marked morphological changes, at least in desmids grown in culture. His study of *Micrasterias* does, however, point to the conclusion that if progress is to be made in desmid taxonomy, then in addition to descriptions of external morphology, some attempt should be made to provide cytological data, for such may provide crucial information. Thus he points out that Krieger (1939) reports that occasionally *M. denticulata*, *M. rotata* and *M. thomasiana* can be confused morphologically. However, if their chromosomes in metaphase can be examined, the three species can be easily distinguished.

Heimans (1969) comments on the taxonomy and interesting geographical distribution of *M. mahabuleshwarensis* var. *wallichii* and the 'very closely related, but clearly different species *M. americana*'. The latter appeared very suddenly, in 1952, in the Netherlands, and Heimans records that Teiling informed him that in a pool in Sweden where *M. mahabuleshwarensis* was for many years alone abundant, *M. americana* suddenly appeared, though in small numbers. Heimans adds that although these two species are closely related, and both are rather variable, he cannot accept 'that the one has originated from the other on the spot!' However, as he also points out, in N. America the two species are so completely connected by numerous intermediate forms and transitions that they have come to be regarded as one very variable species complex (Prescott and Scott 1952). Clearly, cytological studies would be interesting to pursue, both in naturally occurring populations and in laboratory clones raised from such populations (see Blackburn and Tyler 1980).

In attempting to investigate the causes for the considerable range of desmid morphology which can arise in a given species, it would seem clear that a distinction should now be drawn between nature and nurture, nurture having been emphasised to date to the exclusion of possible genetic factors. Carefully planned culture experiments should help to establish which variations are genetically and which environmentally determined (cf. Mix 1965), and of course experience now indicates that for a given species a range of clones should be investigated whose chromosomal constitution is known. It should also be borne in mind that mutability might be increased in culture, and that under laboratory conditions mutations can accumulate and change the appearance of the clone being investigated.

One further point of interest with regard to Kasprik's study of *Micrasterias* is that although many species are very variable, such as *M. americana*, and exhibit quite profound karyological difference, there are those which are morphologically remarkably stable with similar chromosome complements in different clones (e.g. *M. denticulata* var. *angulosa*, *M. papillifera*, *M. rotata* and *M. thomasiana*). These taxa, according to Krieger (1939), are all recorded as reproducing sexually (i.e. their zygospores have been described), but zygospores are unknown in *M. americana*. The inference drawn from this is that sexual reproduction has a stabilizing influence on the karyotype since only very similar chromosomes complements will permit unimpeded pairing at meiosis. Further investigations of the relationship between chromosome complements and the incidence of sexuality, not only in *Micrasterias* but in other desmid genera, would be instructive.

Kasprik questions to what extent knowledge of chromosomes throws any light on the systematic connections which would seem to extend beyond the species in *Micrasterias*—and indeed in other desmid genera e.g. *Staurastrum*. He reports only a partial consistency with respect to the groups distinguished by Krieger (1939) in that the *crux-melitensis*, *apiculata* and *papillifera* groups, the chromosomes would appear to show some uniformity; but not so in the *sol* or *denticulata* groups. Any connection between individual groups and their possible origins are far from solved, and to what extent chromosome studies throw additional light on such presumed relationships remains to be seen.

MEIOSIS IN DESMIDS

Although as long ago as 1899 Klebahn noted the occurrence of two successive nuclear divisions in the germinating zygospores of *Cosmarium* and *Closterium*, their significance was not appreciated, since the process of meiosis was not at that time fully understood. In 1927 Pothoff described, in part, zygospore germination in *Hyalotheca dissiliens*. Fox (1958) gave a brief description of it in *Closterium acerosum* while the metaphase I chromosomes of *Netrium digitus* have been illustrated by Biebel (1964). Brandham and Godward (1965) have provided the most detailed account to date of nuclear and chromosomal events during the meiotic divisions associated with zygospore germination in two varieties of *Cosmarium botrytis*, the homothallic var. *tumidum*, and a heterothallic var. *botrytis*.

In both varieties it is only immediately before zygospore germination that the two nuclei, derived from the parent cells which conjugated to produce the zygospore, fuse. Brandham and Godward believe, since very few cells were seen to contain a fusion nucleus, that this stage is short-lived and that the fusion nucleus passes rapidly into meiotic prophase. Although difficult to observe because of unsatisfactory fixation, in the early stage of prophase the bivalents become visible as a tangled mass of threads which in the var. *botrytis* condense quite rapidly but remain 'fuzzy' and there is no indication of their division into chromatids until diakinesis. The nucleolus, which in the var. *tumidum* is said to be present as a cluster of drops, disappears very soon.

Diakinesis is said to be very obvious and in var. *tumidum* 60–70 bivalents were typically cross-shaped, though a few had two; but the loops between the chiasmata could hardly be observed because the chromosomes were so short and thick. Ring bivalents were not seen. In var. *botrytis* chiasmata were clearly distinguishable, the greatest number in a single bivalent being 3, one terminalized and two interstitial. They differed from those of var. *tumidum* in that they were large enough for the loops between chiasmata to be clearly visible. In these, four chromatids of a cross-bivalent (i.e. one having a single unterminalized chiasma) are clearly demonstrated in one of Brandham and Godward's photomicrographs (their Fig. 16). They comment that this confirms that the chiasmata found in the Zygnemaphyceae are real points of crossing-over and not just 'matrical stickiness' of the type found by John and Lewis (1957) in *Periplaneta*.

At late diakinesis each of the four chromatids present in a cross-bivalent straighten, and in so doing they separate slightly and form a ring. However, they are not joined in pairs as in most organisms, so that it is not possible to distinguish sister chromatids. Brandham and Godward emphasise that this lack of joining between chromatids is evidence for the absence of a localized centromere. Moreover, it confirms King's (1960) conclusion in this respect with regard to the mitotic chromosomes of desmids (see p. 174).

A further point of special interest resulting from observations made on the first meiotic divisions in both varieties of *C. botrytis* is that they are asynchronous. Thus some bivalents may not be fully terminalized, while others, much more advanced in their movements, have separated and the chromatid pairs are already on their way to the poles. The result of this asynchrony is that there is no well defined metaphase plate, the bivalents and chromatid pairs being scattered over a wide area of the spindle. This meiotic spindle differs from the typical truncated mitotic spindle described by King (1960) in that it has tapered poles. Both truncated and tapering spindles were observed at metaphase II.

From squash preparations of metaphase I of var. *botrytis* counts of 11 bivalents and 18 separate chromosomes were obtained ($2n=40$). Brandham and Godward attempted to determine whether the 18 separate chromosomes were univalents that had never paired, or were derived from bivalents that had completely terminalized earlier, so that before the preparation was made they had already divided into their constituent chromosomes. Eleven meiotic stages ranging from mid-prophase to early anaphase I were each scored for the number of bivalents present: 3 were of prophase; 5 early to mid-metaphase I and before the initiation of any anaphase movement; three were of later stages. The number of bivalents from early to late stages of this first meiotic division were, in order: 17, 15, 15, 13, 14, 12, 11, 10, 3, 2, 2. From these figures it can be concluded that possibly some chromosomes do not pair during meiosis and remain as univalents, but it is quite clear that there are more bivalents present at prophase than at metaphase I. Hence it can be deduced that some bivalents have terminalized and their chromosomes separated while other bivalents are still associated and have one or more chiasmata. A certain degree of synchrony in terminalization at the later stage of metaphase I is indicated by the fact that at a later stage the number of undissociated bivalents suddenly drops to 2 or 3.

As Brandham and Godward (1965) emphasize, in most organisms homologous chromosomes separate synchronously at the beginning of anaphase I and that there is no lag phase after separation, the chromosomes moving towards the poles immediately. In *C. botrytis* howeves, and no doubt in other desmids, the first meiotic division differs strikingly from this pattern of behaviour in that bivalent dissociation and anaphase I are both asynchronous; also because there is a distinct lag phase between these two events.

CHAPTER 8
THE MORPHOGENESIS OF THE DESMID
CELL

The credit must go to Waris for recognising the great potential inherent in certain desmid cells, and in particular those of the genus *Micrasterias*, for the elucidation of some of the fundamental problems of cell morphogenesis (Waris 1950a, b, 1951). Thus it was he who realized that, with *Micrasterias*, it might be possible to explore why cells adopt their particular and characteristic shape. He chose this desmid for four reasons. First, its cells are large and in consequence comparatively easy to handle individually; second, they can be readily grown in culture; third, it is possible to follow in detail the development of the complex, but normally symmetrical, lobed morphology in each semicell following cell division (Fig. 72); fourth, they contain a large, clearly visible nucleus which, it was discovered, could be moved about within the cell, with the result that binucleate, or anucleate cells can be produced. Hence these desmids provide unique material for the study of the relative importance of nucleus and cytoplasm in cell differentiation.

Because *Micrasterias* cells are large enough to be seen with a dissecting microscope it is possible, with patience and much practice (and a steady hand), to pick individuals up with a fine glass capillary and to move them about with glass needles. If an experiment demands that cells are to be treated in a particular way during a certain phase of the nuclear cycle, live cells can be examined in a watch glass with a $\times 40$ water immersion objective and again, after considerable experience, it is possible to recognise those nucleolar states which can provide a useful clue to the mitotic state of the nucleus (see p. 158). Once recognised, it is then possible to predict the onset of nuclear division in sufficient time to perform the required experimental treatment at the desired mitotic stage. For example, polyploid, aneuploid, complex or enucleate cells can be produced by centrifugation when the nucleus is in metaphase. However, to achieve this, centrifugation tubes have to be prepared which will hold cells with their axes orientated along the length of the tube (Waris 1950a, Kallio 1951).

In their pioneering studies, Waris and Kallio focussed attention primarily on the external cell architecture, and specifically on problems

185

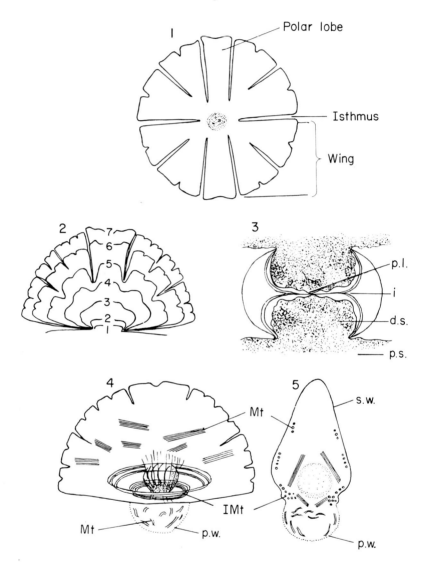

Fig. 72. Morphogenesis in *Micrasterias*. 1 Diagram of a *Micrasterias* cell in front view showing positions of polar lobes and wings relative to the isthmus. 2 Profiles of a semicell during development, the numbers indicate different stages reached at successive intervals of 40 min (after Lacalli). 3 An early stage in the initiation of the polar lobe (p.l.): d.s. = daughter cell; p.s. = parent cell; i = isthmus. 4 Diagram of the systems of microtubules in a developing *Micrasterias* cell in front view. 5 The same in side view. Mt = microtubules; IMt = isthmal band of microtubules; p.w. = primary wall; s.w. = secondary wall.

relating to the growth of the cell wall which leads to the eventual formation of this desmid's characteristic lobes. By applying different types of experimental treatments, such as centrifugation and irradiation which are well summarized in their 1964 paper on morphogenesis, they have demonstrated how cell development can be quite dramatically modified. From the study of these growth aberrations they have drawn valuable conclusions about the possible factors which are active, and therefore significant, during normal cell development.

Each of the two semicells of a fully developed *Micrasterias* cell is characterized by a central polar lobe, on each side of which are the deeply incised, flattened lateral wings (Fig. 72: 1). As indicated earlier (p. 108), during the 4–6 hours following mitosis, the elaborate morphology of the semicell is reproduced in each developing daughter cell by the growth of the septum of the primary wall that forms across the cell isthmus and separates the newly forming cells. The septum is at first a simple, circular plate but soon it becomes hemispherical. Then a central broad protrusion appears on the hemisphere and, by its continued elongation, develops into the polar lobe. While this is taking place lateral protrusions give rise to two wings by their elongation and repeated branching. Since the polar lobe and adjoining lateral wings are restricted to the plane in which the polar lobe and wings of the parent semicell lie, the biradial symmetry of the parent is transmitted to its daughters.

It has already been indicated (p. 176) that change in the cell morphology of desmids is a phenomenon that may reflect nuclear changes, as well as altered environmental conditions. Thus in *Micrasterias*, polyploidy and aneuploidy have been shown (p. 178) to be characterized by the number and arrangements of the lobes. Deficiencies in the ability to produce lobes may be transmitted to offspring through an indefinite succession of cell divisions.

Largely through the work of Waris and Kallio considerable literature now exists relating to the role of the cell nucleus and cytoplasm in determining the morphology of the developing *Micrasterias* semicell. In brief, they have shown that the development of new wings depends on the existence of corresponding old ones, a fact which they interpret as an indication of some plasmatic structural continuity within the cell; however, they also found that nuclear control is needed for cell differentiation (Kallio 1953a, b, 1954, 1957, 1959, 1968, Waris and Kallio 1957). Thus if the nuclear contribution is interfered with, wing lobes will continue to elongate, but lobe branching does not occur (see also Selman 1966).

One of the most interesting results of these studies is that if cells are enucleated at a very early stage of development, at least a complement of three unbranched lobes, corresponding in position to the normal ones, is always produced. This somewhat surprising experimental result led Waris and Kallio to postulate the existence of a cytoplasmic framework with axes responsible for the basic lobed pattern which develops even in enucleate cells. This postulate, coupled with their observations that polyploid cells produce additional wings, or even additional radiation (see p. 179), led them to conclude that the further elaboration of the cell's fundamental threefold pattern is under nuclear control. The existence of species-specific differences and their cytological basis, elaborated in detail by Kasprik (1975, see also p. 178), would seem to reinforce this conclusion, (see also Sano and Ueda 1980).

MICROTUBULES AND THEIR POSSIBLE SIGNIFICANCE

Studies of the ultrastructure of plant cells have provided evidence that cytoplasmic microtubules exert some influence on the establishment of cell form. To what extent they may play a part in the morphogenesis of desmid cells led to a detailed investigation by Kiermayer (1968) of their occurrence and distribution in differentiating cells of *Micrasterias denticulata*. He found that this desmid contained four systems of microtubules (Fig. 72), apart from the spindle apparatus. These are:

1   The isthmus microtubules which surround the nucleus in the cortical cytoplasm of the isthmus.

2   Microtubules in the cortical protoplasm of the old semicell which have rod-like cross-bridges between individual microtubules.

3   Clusters of microtubules adjacent to the post-telophase nucleus, some of which are separated by intertubular structures.

4   Microtubules in the internal and cortical protoplasm of the growing semicell.

Kiermayer (1970) questions whether the microtubules are functional during morphogenesis, since treatment with colchicine and vinblastine, which are known to disorganise microtubules, exerted no influence on the development of shape in *Micrasterias*. Since these substances, however, affected the post-mitotic migration of the nucleus and the chloroplast (see p. 115) he reasoned that the microtubules may play a vital part in controlling the position of the nucleus and the chloroplast (see also Tippett and Pickett-Heaps 1974 and p. 107). They may also be involved in the transport of other cell organelles into the enlarging new semicell, in the subsequent ordering of the organelles, and in anchoring the nucleus in its normal site (Kiermayer 1972).

Annular microtubules, similar to the isthmus microtubules, have been found in other members of the Zygnemaphyceae, and according to

Pickett-Heaps (1975) are known to reappear in the isthmus soon after cell division. Kiermayer (1970b) has proposed that they may play a part in determining the position of the septum (see p. 112) but according to Lacalli (1973), do not seem to be involved with septum growth (see p. 112). However, Lehtonen (1977) observed that one-day treatment with $10^{-6}\%$ trifluraline, and three-day treatment with 1.0% colchicine dissolved the isthmus microtubules, prevented septum formation and markedly affected the structure of the resulting daughter semicell. He recognises that these drugs may in themselves have had direct morphogenetic affects but his observations would rather seem to indicate that the disturbances caused are due to microtubule dissolution, with the result that the basic determination of structure was disturbed. This has been found to occur during mitosis and is connected with cell wall formation in the isthmus. As indicated elsewhere (p. 112), the girdle is formed of primary wall material in early mitosis precisely in the area of the isthmus microtubules. Lehtonen argues that it seems probable that these affect girdle formation by controlling the orientation of the cellulose microfibrils. Robinson, Grimm and Sachs (1976), in studying microfibril orientation following colchicine treatment, have demonstrated that cortical microfibrils have this ability, and it is suggested that the location and orientation of microfibrils may determine the basic polarization of growth. Hence, the disturbance in the orientation of microtubules resulting from the dissolution of the isthmus microtubules will, according to Lehtonen, cause development to be aberrant. This view is supported by the fact that cellulase treatment caused significant modification to the form of the developing semicell when the enzyme was applied during girdle formation, but when applied later, it affected only the size of the semicell.

In addition to the arrangement of a system of microtubules in the early stages of growth and differentiation in *Micrasterias* cells, Kiermayer (1970b), in a further EM study, has shown that two different types of vesicles are 'pinched off' by cisternae of the dictyosomes. These he has termed 'large vesicles' since they range in length from 0.3–1.0 μm, as compared with electron-dense 'dark vesicles' of 80–250 nm. He suggests that the large vesicles may be associated with mucilage production by way of the pores of the secondary wall (p. 94), in contrast to the dark vesicles which are concerned with wall production. Kiermayer also found 'coated vesicles' and 'rod-containing vesicles' in the cytoplasm, the former frequently being fused with the plasmalemma. He states that dictyosomes and cisternae of the rough endoplasmic reticulum are the most prominent cytoplasmic components of the developing semicell and finds them to be clustered round the post-telophase nucleus, along with mitochondria and microtubules, thus indicating a plasmatic stratification during early development.

However, as nuclear migration progresses deeper into the enlarging daughter semicell (p. 115 and Fig. 51), all these organelles are distributed, apparently at random, throughout the cytoplasm.

Because he could find no significance to the position of the systems of microtubules mentioned above to the typically symmetrical pattern of the developing semicell, Kiermayer has speculated that the plasma membrane may maintain a specific pattern in its molecular structure, thereby controlling cell wall material deposition. Thus he suggests that vesicle membranes exhibit particular affinities to local areas of the plasma membrane and discharge their contents as materials for the growing wall only at these loci. This mechanism, he points out, could lead to a patterned growth of the cell wall, despite the observed random distribution of organelles and vesicles and the frequently observed, very vigorous cytoplasmic streaming (cf. Lacalli 1975 and p. 192). Tippett and Prickett-Heaps (1974) support Keirmayer's views, but suggest that as proteins may have a morphogenetic function, the maintenance of their synthesis may be an additional crucial activity.

THE GOLGI APPARATUS

Ueda and Noguchi (1976) have examined the transformations that occur in the Golgi apparatus during the cell cycle of *M. americana* (see also Dobberstein and Kiermayer 1972). Their EM micrographs show that in non-dividing cells the dictyosomes contain 11 cisternae, the most distal of which are reticulate. In these non-growing cells, the Golgi apparatus is found near the surface of the chloroplasts with the cisternal plates parallel to the surface of the chloroplasts. There is rough ER between the Golgi apparatus and the chloroplast. During the development of the new semicell, following the mitosis of the nucleus, the dictyosomes produce electron-dense 'dark vesicles' corresponding to those described by Kiermayer (1970, see above) and Ueda (1972), as well as 'large vesicles', both being formed at the periphery of the distal cisternae.

At the metaphase of mitosis, when the ingrowing septum is initiated, several dictyosomes appear around the nucleus and the ER is located nearby. Also at this stage many dark vesicles can be found at the margins of the cisternae. Within 5 hours after the initiation of the septum, the enlarging new semicell has attained the same size as the parent semicell. Although by this time the downward growing chloroplast has not fully expanded, about 90% of the dictyosomes occur in the new semicells, especially near the new semicell wall and the remainder are still in the parent semicell. The transformations which occur in the Golgi apparatus as they accompany successive stages in the developing semicell are clearly illustrated in Ueda and Noguchi's Fig. 20.

The phosphatase activity, especially of the Golgi apparatus, has been studied by Noguchi (1976), in both resting and developing cells of *M. americana*.

THE GROWTH OF POLAR AND WING LOBES IN MICRASTERIAS

The major problem still to be solved is how the various cytoplasmic and nuclear determinants act to produce the remarkably complex form of the mature *Micrasterias*, or any other desmid semicell. In his paper on the asymmetry of desmid cells Teiling (1950), in discussing the deviations that can occur, (see p. 32) suggested the occurrence of 'meristematic organelles' within the cytoplasm. He suggested that during the development of each semicell from 'the primary globoid outbulging', its morphogenesis is directed by changes in the activity of these organelles by a coordinating centre. He suggested that each process and emergence must be the result of a localised growth, whose duration and intensity is also determined by its activity. Much later, support for this notion came from the demonstrations by Kiermayer and Jarosch (1962) and Kiermayer (1964), who demonstrated patterns of differential wall deposition and also in the strength of membrane attachment to the cell wall in new semicells. As mentioned previously (p. 110), Kiermayer thinks that the septum is particularly significant in respect of semicell pattern and postulates that it might act as a morphogenetic template (Kiermayer 1967, 1970a).

Kallio and Lehtonen (1973) have shown that the potential for axis formation in *Micrasterias* is transmitted from the polar lobe and wings of mature semicell to its developing daughter by way of 'growth centres' which are not under the control of the nucleus, and which would seem to be situated at, or near, the isthmus. That such growth centres exist comes from experiments in which there has been a development of 3-lobed cells following the removal of the nucleus by centrifugation, or by the elimination of nuclear control during mitosis with UV radiation or antibiotics (Kallio 1951, 1957, 1968, Selman 1966, 1972). These studies have demonstrated that the basic symmetry of the semicell is determined very early in its development and that its fundamental structural units form without continuous nuclear control, although there is no subsequent semicell differentiation. As pointed out by Lehtonen (1977), the formation of growth centres presupposes the existence of a plasmatic template. If, as he states, such templates are present in the cell wings, their disturbance may be expected to cause structural changes in following generations. In his study of morphogenesis in *M. torreyi* and *M. thomasiana*, using UV microbeam irradiation and chemical treatment, such changes were produced when a wing was irradiated during cell division. He found the first daughter semi-cell to be smaller and less detailed in its structure than an untreated one. In the next generation, however, the corresponding wing was in some instances completely absent (see his Fig. 2J, K and L). That irradiation did not change the basic structure of the semicell immediately suggested to him that the structure of the daughter semicell had already been laid down before this treatment, a result explained by the theory of growth centres.

Lehtonen has produced additional evidence for the existence and significance of growth centres by subjecting the isthmus to partial irradiation. Thus, one side of the lengthening isthmus, especially the wall, was irradiated during, or soon after, mitosis. To ensure that the effect was not on the nucleus, the treatment was given with the nucleus in its normal site, and also to avoid its possible damage, by its removal from the isthmus with trifluraline treatment. This treatment of the isthmus appeared to affect the postulated growth centres directly, causing immediate structural change. From such evidence Lehtonen concludes that the isthmus is probably the site where the growth centres form and that they must be organised in this crucial region of the daughter cell wall before the morphogenesis of the daughter semicells begin.

Lacalli (1975), using a laser microbeam from a helium–neon gas laser (Lacalli and Acton 1972) to irradiate dividing cells of *M. rotata*, showed that only certain regions of the developing primary wall are required for continous growth and morphogenesis and that much of the surface plays only a passive role. Lasing the septum at a point exactly between the two developing semicells prevented the formation of the polar lobe in both daughter cells without causing much alteration to the wings. Damage was most conspicuous in lasings of semicells sufficiently developed to have major lobes and corresponding notches apparent. Such treatment caused portions of the final complement of lobes to be absent. From his study Lacalli concluded that specific regions of the cell surface, the lobe-tips, play a much more significant role in morphogenesis than other regions of the cell surface. He also demonstrated that the lobes grow and develop independently of one another because one lobe is not adversely affected by a damaged neighbouring lobe; also it does not compensate for an increase in its own growth when an adjoining lobe is missing. He then goes on to argue that it should be possible to map the cell surface in each stage in development 'to determine the position and size of the sensitive regions and, through a series of stages, to trace the developmental history'. Although attempted, this proved, for a variety of reasons, impossible to achieve. That the formation of new wall material is concentrated at the tips of the lobes was revealed by Lacalli in autoradiograms showing the incor-

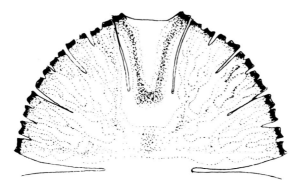

Fig. 73. The veining pattern of methionine label for primary cell wall ghosts. Regions of dense and uniform labelling occur at tips of lobes (black areas); veined labelling (stippled) (after Lacalli).

poration of methyl [³H]methionine and C-I-[³H]glucose (Fig. 73).

The specially endowed area of tip wall, the growing point, or as Lacalli (1975a) terms it the 'singularity', must be responsible for the observed growth of the lobe tips. These presumed, highly structured sites, which he indicates can be no larger than 4–5 μm diameter, though they could be much smaller, must be initiated in the cell wall with the organisation of Lehtonen's postulated growth centres. As he states, since their partial damage is possible, 'they apparently comprise larger areas, and growth is then gradually concentrated in smaller areas, which become the tips of the primary lobes'. Undirected or undetermined growth, according to Green (1969), invariably produces a spherical structure. The UV irradiation of *Micrasterias* cells can disturb their differentiation by causing the destruction of functioning growth centres, but it does not prevent the formation of the cell wall, or at least the so-called bulge wall, and after such treatment spherical daughter semicells can result.

Although the intial form of the developing semicell in *M. rotata* has been shown by Lacalli (1975a) and Lehtonen (1977) to involve growth centres and 'singularities', its final form must result from a combination of 2 additional processes, branching and lobe broadening. These aspects of *Micrasterias* morphogenesis have been studied in *M. rotata* and especially *M. radiata*, also by Lacalli (1975b). From this study information about the participation of specific regions of the cell wall in cell development was obtained by irradiating cells at various stages of development with his laser microbeam (see p. 192) after treatment with Alcian blue. Autoradiograms were prepared to reveal patterns of cell wall incorporation, and of cell wall ghosts from cells left for varying times during their development in culture solutions containing [³H]glucose.

These autoradiogram studies showed that although incorporation was concentrated at the lobe tips (p. 192), there was, in addition, a secondary pattern on the semicell surfaces which Lacalli compared to the veining in a leaf. In general, the so-called veins followed the midline of wing lobes, branching wherever lobes branched, and become much wider at the branch points (Fig. 73). There was, however, always a band of unlabelled wall encircling the polar lobe at its base, as were also areas in the region of the lobe notches. Lacalli emphasises the marked difference in the veining of the polar lobe from that of the wings, for the midline of the polar lobe is also unlabelled. Veins, instead, form a V down either side of the midline to unite at the base of the polar lobe. He also stresses the special response of the polar lobe to laser irradiation in that it could be inhibited from developing by lasing the central point of contact between daughter semicells. Occasionally, however, treatment intended to do this produced, rather, a duplication instead of an elimination of the polar lobe. This never happened, however, when wing lobes were lasered, though they were always eliminated when wing lobe tips were irradiated.

In the production of cylinder-like polar lobes, typified especially by those of *M. radiata*, the process of tip growth, according to Lacalli, is responsible for their formation. However, as soon as branching occurs, 'growth of a fundamentally different type has been imposed, because constraints of geometry prevent branches from forming, unless additional insertions of wall material are allowed.' The geometry involved in the branching by such cylindrical lobes is clearly illustrated in Lacalli's Fig. 8 (1975b), in which it is emphasised that new lobes with flat bases cannot be set on top of the original cylinder without spaces being left in the wall. Since, however, the cell wall is formed continuously 'without leaving sur-face gaps, later to be filled in, it is perhaps more accurate to say that the process of tip branching must be accompanied by cell wall growth else-where than at the lobe tip that compensates geometrically for the change from one growth site to two'.

Also from his autoradiographic studies, Lacalli has provided evidence consistent with the view that regions of the wall must be filled in, or expanded, by additional synthetic activity. In the resulting pattern of wing extension, labelling has been shown to occur at the points at which the lobes have branched and where his analysis demands the insertion of additional wall material.

With reference to the process of lobe broadening, Lacalli's autoradio-grams show that a narrow, continuous band of wall incorporation can be traced along the midline of the semicell lobes to connect up with the wider regions. This is to be expected if a broadening of the wings occurs by the insertion of new wall material along the midline of the lobes as postulated by his geometrical analysis of growth.

Finally, Lacalli stresses that branching and broadening are inextricably associated processes, one depending on the other, and that both are possibly responses to an additional underlying process. Thus he postulates that growth sites (he terms them singularities) which determine tip growth, though discrete structures, can be duplicated and that when this happens, the resulting stresses within the wall would produce an insertion of essential additional wall material in a veined pattern.

CHAPTER 9
THE ECOLOGY OF DESMIDS

INTRODUCTION

The major aims confronting the desmid ecologist would seem to be four-fold:

1   to investigate the distribution of desmids, and discover the types of habitats in which they live;

2   to explore their spatial and temporal distribution in these habitats;

3   to examine the presumed multiplicity of factors involved in the control of this distribution;

and since, in the final analysis, ecology is concerned with the study of the structure and functioning of ecosystems (Odum 1971),

4   to attempt to discover the role that desmids play in the ecosystems in which they are found.

OCCURRENCE (SPATIAL DISTRIBUTION)

Desmids are almost exclusively freshwater algae confined in their distribution to natural waters characterised by low salinities and hence low specific conductivities. Only a very few species have been found in waters of high ionic content. Thus Grönblad (1956) recorded *Closterium incurvum*, *Cosmarium granatum* and *C. humile* in brackish water with a salinity of 4.2% in the eastern U.S.; he also found *C. granatum* in Finland (1953) in brackish water which was heavily polluted with sewage and with a salinity of 2.36%. Förster (1972) also found desmids in the carbonate-sulphate Venezuelan Lake Valencia with a salt concentration of 0.6%. Nygaard (1976) reports (with some reservations) the discovery of several *Cosmarium* species, and again *Closterium incurvum*, from an Arctic salt lake in West Greenland. This lake was remarkable for its high bicarbonate and chloride and extremely low sulphate concentrations. Bocher (1949), who made the collections and carried out the chemical analyses, comments that the pH was 8–9, and the concentration of chloride ions in the lake was 169 mg

Cl$^-$/1 and 65 mg Na$^+$/1. The only experimental study on the effect of Cl$^-$ and Na$^+$ ions on the growth of a desmid (*Netrium digitus*) would seem to be that of Hosiaisluoma (1976) summarized on p. 225 (see however, Hofler 1951).

In the 1920's and 30's when great confidence (but apparently little understanding of why or what was being measured) was attached to the distribution of organisms and the prevailing pH, numerous lists were published attempting to correlate desmid distribution with pH. The main conclusion to be drawn from such surveys was that by far the greatest number of desmid species occur in acid waters (pH 4.0–7.0) which hence were of low alkalinity (soft waters). Recent research, while not disproving this fact, has however shown that many species occur, often in considerable abundance, in waters which are markedly alkaline (hard waters) (Bland and Brook 1975, Brook 1965, Moss 1973a, Woekerling 1976, Gough and Woekerling 1976, Woekerling and Gough 1976, Gough 1977). Indeed, species diversity is highest in acid waters and *Sphagnum* bog pools are desmid habitats 'par excellence', especially it seems in the temperate zone of the N. Hemisphere (see, however, p. 200). Here they may occur in gelatinous masses, floating freely or adhering to submerged aquatics such as *Utricularia* species. A somewhat similar habitat to the latter, and also qualitively and sometimes quantitatively rich in desmids, is to be found in the interstices of the submerged vegetation of larger bodies of water such as ponds and lakes. They may also be abundant on the felts of filamentous algae which may develop on and around these plants. Some desmids occupying such habitats would seem to be fairly closely bound (though rarely attached) to the macrophytes with which they are associated, while a considerable number appear to lead an almost free-floating existence in the water surrounding the plants, so that quite frequently they are carried into the open water of the ponds and lakes in which they occur. This diverse community has been termed the 'metaphyton' (Behre 1956). In addition to such, often rich, desmid habitats, these algae can be a conspicuous component of the phytoplankton community of the open water of ponds and lakes, especially in calcium-poor, soft-water lakes. Griffiths (1928) proposed three terms to describe these major desmid habitats, using 'terraqueous' for the community in bogs, small ponds or ditches, 'bentho-planktonic' for that amongst the submerged macrophytes of the littoral, and 'limnoplanktonic' for the open-water community of larger bodies of water.

In the introduction to the Desmid volume of Rabenhorst's Kryptogamenflora, Krieger (1933) lists the desmid communities described by numerous authors, though, as Heimans (1969) has emphasized, the same community is recorded, without distinction, sometimes as an association, then as a formation or again as type, in the sense of a flora-element. It would be better, he urges, if the original definition of the concept of 'association' as defined at the Third International Botanical Congress (1910) was used. According to that resolution, an association is a plant community characterised by its own typical combination of species. However, even this term may be difficult to apply to unattached, free-floating organisms such as desmids, and so he suggests when studying local combinations of species, their ecology and range of distribution it would be best 'to completely avoid the use of such terms as association, formation, element and replace them when necessary by a periphrasis'! (Heimans 1969, pp. 58–59)— presumably meaning 'some other term'?

The most recent approach to the study of the structure and pattern of desmid communities is that by Péterfi (1974). In this, the phytosociological and ecological affinities of the most commonly occurring desmids in some Rumanian peat bogs were tested by sign-correlation analysis of his desmid presence and absence data. His results indicate the existence of two clusters of positively associated species. The first contains desmids with positive affinity and which occur in oligotrophic habitats (pH 4.0–4.5). A second group of positively associated species are characteristic for less acidic (mesotrophic) sites (pH 5.0–6.5). Several species, however, were found to form an intermediate group showing positive significant correlation with either the first or second groups. Hence, Péterfi concludes that their distribution would not appear to be strictly limited by ecological preferences, since they occur in both habitats.

The structure and pattern of the oligotrophic desmid communities were also examined in terms of dimensional species relationship and the strongest positive association was obtained between *Staurastrum spinosum Actinotaenium cucurbita, Netrium oblongum, Cylindrocystis crassa, Bambusina brebissonii* and *S. quadrispinatum*. The arrangement of these six species in the cluster diagram indicates that they constitute what Péterfi designates the 'central' structure of the community type. He suggests that the presence of one of these central species occurring with considerable frequency in any locality presumes that at least some, or even all, of the remaining species of this cluster should be present. Such association analysis would seem to be a method which should be further explored for the elucidation of the phytosociological and ecological affinities between desmid species and various environmental factors. By such means it is hoped that the general pattern and structure of particular communities can possibly be more precisely defined (see also Palamar-Mordvintseva 1978a).

In collections of plankton from the open water of a lake it is possible to find desmids typical of each of the three habitats proposed by Griffiths (Brook 1959a, Duthie 1965, Bland and Brook 1975). Thus, even forms from terraqueous habitats can survive, though they may not reproduce in the plankton. In fact it seems to be recognized that planktonic desmids have originated from, and are morphological adaptations of, terraqueous species from bogs or marshes surrounding the lake, or they are benthoplanktonic, derived from the submerged weeds of the littoral (Wesenberg-Lund 1905, West and West 1909, Griffiths 1928). After examining in detail the plankton of 98 large, deep Scottish freshwater lochs, and assuming that any desmid frequently found undergoing cell division in the open water of these lakes was planktonic, Brook (1959) drew up a list of euplanktonic desmids (Table 9). He then argued that the same taxa present in the plankton of smaller, shallower lakes can also be considered to be euplanktonic. Also listed (Table 9) are species which, although found most often in boggy habitats, have also been found dividing in open water and thus can be considered to be facultatively planktonic. Brook provides some evidence to indicate that the plasticity of the desmid cell allows the rapid adaptation of terraqueous desmids to the planktonic habitat by the establishment of ecological forms with different morphologies and particularly a reduction in their ornamentation (Brook 1959a, Figs 3–6 and Brook 1959b). Such an adaptation, because of the considerable plasticity of the desmid cell (p. 32), may take place annually. The contrary view (West and West 1909) is that the process of adaptation has been gradual and evolutionary (genetic) and has been accompanied by the development of copious mucilaginous sheaths as aids to floatation. West and West discredited a chance or yearly adaptation, stating that neither planktonic desmids, nor any other, undergo any seasonal form variation (see, however, p. 204), or changes in buoyancy, and that desmid plankton overwinters in the open water.

Table 9. List of euplanktonic desmids from the British freshwater phytoplankton (Brook 1959).

Closterium setaceum	*Xanthidium antilopaeum* var.
Cosmarium abbreviatum	*hebridarum*
var. *planctonicum*	*X. subhastiferum* var. *murrayi*
Micrasterias mahabuleshwarensis	*Staurastrum cingulum*
f. *wallichii*	*S. cingulum* var. *obesum*
Staurodesmus aristiferus var. *gracile*	*S. cingulum* var. *affine*
S. curvatus	*S. boreale* var. *planctonicum*
S. cuspidatus var. *canadense*	*S. pingue*
S. indentatus	*S. pseudopelagicum*
S. jaculiferus	*S. lunatum* var. *plantonicum*
S. megacanthus var. *scoticus*	*S. anatinum* f. *longibrachiatum*
S. sellatus	f. *pelagicum*
S. subtriangularis	f. *paradoxum*
S. triangularis	f. *glabrum*
Euastrum verrucosum	*S. longipes*
var. *reductum*	*S. furcigerum* var. *armigerum*
Spondylosium planum	*S. longispinum*

O

In an attempt to throw further light on the status of desmids in the plankton Duthie (1965) studied desmid populations in a shallow, oligo-trophic Welsh lake, Llyn Ogwen. He attempted to answer such questions as to whether, in spite of the lake's shallowness (average depth 2.1 m), a distinct community of planktonic desmids could be recognised?; where do desmids overwinter?; do they all multiply in open water if transported there?; can any seasonal changes be detected in cell morphology?

His results indicate that while no difference in behaviour could be recognised between representatives of Brook's euplanktonic and faculta-tive plankters, a real difference did exist between these two presumed categories and the clearly non-planktonic species (Duthie's Tables 3 and 4), these being principally benthic in habit and rarely found in the plankton. Two species, *Staurastrum ophiura* and *S. tohopekaligense* var. *trifurcatum*, not included in Brook's euplanktonic taxa, were common in the plankton but rare in sediments and inflows. Since there were no weed beds to act as a littoral source for them (see below) their occurrence suggested to Duthie that they were independent of the benthos, and truly planktonic. However, some other 25 species and varieties that showed planktonic tendencies appeared to form a distinct community but were able to multiply on both the sediment and in the plankton. Duthie also mentions that several species constituting his planktonic community were found in nearby bogs or bog pools. Round's (1953) observations on desmids found on the peaty sedi-ments of the base-rich Malham Tarn are of interest here, as is Nauman's (1927) view that the ecology of phytoplankton organisms in general varies considerably with the environment. Thus he states that algae living on sediments of lakes poor in Ca, N and P, tend to occur in the plankton in more eutrophic lakes. However, as Duthie (1965) points out, there is as yet no clear support for this in the literature with respect to desmids.

Another aspect of the problem of the status of the desmid plankton has been investigated by Bland and Brook (1974) in a study of the spatial distribution of desmids in lakes in N. Minnesota, U.S.A. In this study the relationship between desmid populations in the littoral, especially those associated with aquatic macrophytes, and those in the open water (the limnetic zone) of the lakes was examined quantitatively. Twelve different species of macrophytes in 10 different lakes were sampled. Some 250 species and varieties of desmids were identified (Meyer and Brook 1969), the majority being associated with submerged aquatics as part of the metaphyton. The relative abundance of desmids was strikingly different on different macrophytes and the differences would seem to be accounted for by marked differences in the underwater morphology of these phanero-gams. Thus *Brasenia*, *Nymphaea* and *Nuphar* which lack underwater branching or dissected leaves supported few, or in the case of *Brasenia*, no desmids. The largest and most diverse desmid populations were on macrophytes with finely dissected leaves such as *Ceratophyllum*, *Myriophyllum* and especially *Utricularia*, and also two *Potamogeton* species which branch profusely. Similar findings, especially with reference to *Utricularia* as an excellent habitat for desmids (and even better than *Sphagnum*), are those of Woelkerling (1976). It has been suggested that the richness of *Sphagnum* as a desmid habitat may, at least in part, be due to the ability of the moss to act as an ion-exchange system thereby reducing the pH of the surrounding water (Rose 1953, Bell 1959). This would result in an increase in the availability of free CO_2 and provide conditions favourable to many desmids (Moss 1973a, c). Why *Utricularia* should have such an abundant, and according to Bland and Brook (1974) and Woelkerling (1976), very much more diverse desmid assemblage than do stands of *Sphagnum* is clearly a topic worthy of further investigation. However, some comments of Heimans (1969, p. 75) with respect to this *Utricularia*-desmid association would seem to be worth quoting, 'Not only does a floating vegetation of *Utricularia* provide a particularly suitable development milieu for many desmids, but it is also an excellent means of transport, so that a pool where *Utricularia* is brought, possibly by waterbirds, can offer an exceptionally favourable accessibility for Desmids'. The term 'accessibility' was first proposed by Heimans (1934, 1954), and implies that just as climatic, edaphic and biotic factors only permit the occurrence of specific taxa to survive and reproduce in habitats in which their particular requirements are satisfied, so that the 'accessibility factors' will sift out the species according to their abilities to be transported and disseminated (see p. 208). Heimans (1969) points out with regard to the geographical distribution of desmid species, their ubiquity must first be related to the time available since their origin; secondly to the physical, chemical and biotic conditions of a particular environment; thirdly to the accessibility of that environment.

A further finding of interest by Bland and Brook was that there were significant differences between populations of desmids associated with the same species of macrophytes occupying different regions of the same lake. Thus counts from 25 standard samples of *Potamogeton robbinsii* from an area in one lake protected from wind, and hence wave action, yielded 2473 desmid cells, while the same number of samples from an exposed area yielded only 94 cells. It was also found that in samples from the protected site 12% of desmid cells were actively dividing and only 4% appeared to be dead or moribund, while from the macrophytes in the exposed area, 56% of the desmids appeared to be dead or dying and none were seen to be dividing.

With regard to the overwintering of desmids, at least in the Minnesota lakes this appeared to be related to the form of the winter buds produced by the macrophytes. Thus both *Potamogeton robbinsii* and *P. amplifolius* produce large winter buds and the plants themselves remain for the most part intact under ice cover and continue to harbour considerable desmid populations. Other macrophytes degenerate in winter and their remains contribute to the sediments and so presumably do their associated desmid populations. The potamogetons not only provide an overwintering habitat but since these plants tend to break up during the spring overturn, their large buds, which also contain some desmids, are carried by currents to various parts of the lake and there to be the seeds of new populations of the metaphyton in spring-growing weedbeds. From such observations it seems clear that the presence, or absence, of particular macrophytes may be of very great importance in determining the diversity and abundance of the desmid flora of a given lake.

As to the distinction between truly planktonic (euplanktonic) desmids and those present in the littoral, but occasionally carried into the limnetic regions as chance or tychoplankters (Duthie's Group 4), Bland and Brook's study would seem to confirm Duthie's findings in Llyn Ogwen. Thus they found that euplanktonic species in one lake may be tychoplankters in another, or even in different areas of the same lake. Indeed, they conclude that whether a desmid is planktonic or not seems to depend on certain physical and biological characteristics of the particular lake they inhabit. They support this conclusion especially from observations made of differences in the desmid populations between a well protected and an exposed region of the same lake (Squaw Lake), each of which had earlier been found to support markedly different phytoplankton populations (Baker and Brook 1971). That these were associated with differences in the water circulation patterns between the two areas was clearly demonstrated by time-depth diagrams which showed for example that the bottom 2 m of the protected area were anoxic from mid-June until the autumnal overturn in mid-October while the exposed area never became anoxic. In the protected region of the lake the analysis of a series of depth-profile transects showed that desmid abundance declines from up to 150 per 10 ml in the littoral to none in the open water, while in the transect taken in the exposed area, though only a maximum of 14 per 10 ml were present in the littoral a large number of the open water (limnetic) samples contained desmids (1–13/10 ml). Thus where water circulation is poor it would seem that desmids sediment rapidly and are not carried into the open water. This result was confirmed in an adjacent lake where similar trends in distribution were observed. Clearly Bland and Brook's results support the hypothesis that in lakes, or regions of lakes with good water circulation there will tend to be a greater number and greater diversity of planktonic desmids than in those in which there is less active or poor circulation.

This was further supported by an examination of lake morphometry as indicated by relative depth (Hutchinson 1957, Baker and Brook 1971) which tends to reflect circulation in relation to the size range of the desmid cells found in their plankton. Five of the lakes so examined were comparatively large, dimictic with relative depths between 0.57 and 2.72 m, while four and the protected area of Squaw Lake (see above) had been shown to be meromictic (Baker 1973) with relative depths from 5.27 to 7.92 m. Analysis of the data revealed a positive relationship between relative lake depth and the size of the desmid cells present in their plankton. Thus in the dimictic lakes cells ranged in size from 9–300 μm, while in those with high relative depths, and tending to meromixis, the range was only 9–40 μm.

Bland and Brook also related the total numbers of desmids present in mid-summer in the limnetic areas of the 10 lakes they studied and made the somewhat surprising discovery that small desmids were much more numerous in the meromictic lakes than in the larger ones. Thus in three of the lakes *Staurastrum tetracerum* and *Staurodesmus extensus* contributed more than 82% of the total desmid populations. One interesting observation relating to the vertical distribution of such small desmids was that in two of the meromictic lakes they were most numerous in the epilimnion, while in a third they occurred in the hypolimnion. It was considered that this difference in vertical distribution was related to the fact that in the first two lakes large metalimnetic populations of *Oscillatoria redeckii* and *O. agardii* var. *isothrix* persisted, so that below 6 m there was virtually no light penetration, whereas in the third, although significant metalimnetic populations of cryptomonads and *O. ornata* were present, their densities were not as great as the other metalimnetic populations. Hence the photic zone in the third lake was much deeper.

All of these observations would seem to indicate that water circulation may be a factor of great, and at times overriding, importance in determining the nature and abundance of the desmid populations of lakes. Hence in small lakes circulation in the littoral may not be so intense as in large lakes, and larger desmids originating either from the sediments or submerged macrophytes are less frequently carried into limnetic regions. In big lakes, because of more active water circulation, larger, and in many cases a greater variety of, desmid species are transported into the open water. Thus in lakes and ponds which support desmids in their plankton there is undoubtedly an annual recruitment of them from sediments and littoral. In many large lakes which do not freeze over in winter, so that continuous water circulation is maintained, desmids have been found to occur in their plankton throughout the year (Brook 1965, Lund 1971, see also p. 208).

Even in a small bog pool (open water about 200 m² and from 5–30 cm deep) Duthie (1965) found that the plankton of the bog water, with the exception of the desmids *Staurastrum margaritaceum* and *Staurodesmus apiculatus*, was dependent on the abundance of algae in the sediments. A period of rain, or a thaw after snow, drastically reduced the plankton which was later replaced from the sediments. As found by Bland and Brook (1974, see also Brook and Woodward 1955), Duthie's observations would seem to suggest that physical factors such as rainfall and wash-outs produce turbulence which may dramatically affect plankton composition.

The effect of wind, and ensuing wave action, on the dispersal of desmids from littoral regions is in agreement with proposals made by both Pearsall (1924) and Fritsch (1931). It is also in accord with Stokes' Law which relates to the free fall of small spheres in a liquid medium. Application of this physical principle, which indicates that the free fall of small spheres varies directly with the square of their radius, suggests that in Bland and Brook's poorly circulating lakes the successful survival in the plankton of small desmids (9–20 μm diameter) was due to their slow rate of sedimentation. Desmids of this size should fall approximately a hundred times more slowly than those 100–200 μm diameter. Duthie (1965) showed, as a result of some brief experimental studies on sinking rates, remarkably good agreement with the recognized ecological status of the species examined. Thus *Cosmarium botrytis* and *Micrasterias truncata*, which are rarely or never found in the phytoplankton, had the most rapid sinking rates; they also had very thin mucilage sheaths. *Staurastrum tohopekaligense* var. *trifurcatum*, *S. furcigerum* and *Cosmarium subtumidum* all had wide mucilage sheaths and lower sinking rates. Duthie attempted to measure sinking rates directly but with little success. Indeed he places reliability on only one figure of 15–20 μm/s which relates to the sinking rate of cultured *Cosmarium botrytis*. This compares with 2 μm/s for *Asterionella formosa* and 10 μm/s for *Melosira italica* (Lund 1959). According to other data gathered by Duthie (see his Table 6) *C. botrytis* has a sinking rate at least 6 times greater than that of the desmid which he found to be a euplankter in Llyn Ogwen, *Staurastrum furcigerum*, whose sinking rate he therefore concludes must be of the order of 3 μm/s.

Both Teiling (1947) and Brook (1959) in formulating possible evolutionary trends in species groups of *Staurastrum* observed elongation of body and processes and a reduction in cell wall ornamentation as an accompaniment to increasing planktonic adaptation. Experiments designed by Duthie (1965) to attempt to answer the question whether seasonal changes in cell morphology could be detected showed that the extent to which a cell became fully differentiated was exceeded by the rate of division, i.e. the cell divided before attaining its full morphological potential.

Clearly in rapidly dividing populations both semicells may not exhibit full maturity (see pp. 108–117) and if this takes place, as it certainly does in natural populations as growth conditions change seasonally, there is clearly support for Brook's suggestion that because of the great plasticity of desmid cells the production of appropriate ecological forms could be a rapid, and seasonal, process. Planktonic species grown in culture (Brook, personal observations) often show a reduction in the length of their processes and an increase in their ornamentation. Hence, for example, *Staurastrum pingue* adopts the appearance of the benthic or metaphytic *S. creulatum* (cf. Brook 1959, Plate XVII Figs 1–6 and 9–11; see also illustrations in the great variation and ornamentation in *S. anatinum*, Plates I–IV in Brook 1959). Reynolds (1940) in his study of seasonal variation in *S. chaetoceras* found that in the summer population the triradiate forms predominate over biradiate facies which are most abundant in autumn, winter and spring; he also found, in support of Brook's thesis, that the arms of biradiate facies vary in a seasonal manner, the length being greatest in June but that they became 'shorter as winter approached, the length increasing in spring and early summer'. It would be most illuminating to examine the sinking rates of these different ecological forms of *S. chaetoceras* in various water temperatures and hence in different water densities (cf., however, p. 237).

An increase in process length and also the extent of a cell's mucilage envelope (Ruttner 1953, his Figs 40 and 41) is generally considered to confer on desmid cells increased resistance to sinking and so is presumed to be an adaptation to a planktonic existence. Duthie's simple geometric figure representing a rotating body with processes in a flowing liquid shows that depending on how a current impinges on the processes placed laterally to it, so the cell will curve either to the right or left. When Duthie measured sinking rates he observed that cells fell haphazardly, so creating the necessary motion. As he states 'in a lake a desmid with processes might behave in a similar manner and instead of being carried along passively, or sinking directly, would always tend to curl away, so greatly increasing its chances of entering favourable currents'. He also comments that his sinking rate experiments took no account of current upon flotation but suggests that an application of Magnus and Benoult's equation (Goldstein 1952, p. 82) might apply to the flotation, at least of *Staurastrum* species.

DESMIDS IN FLOWING WATERS

Some species, again very restricted in number, are found in running water in streams, rivers and possibly in springs, where concentrations of CO_2

tend to be high. As in ponds and lakes such desmids are usually associated with aquatic macrophytes though a few species may be benthic. Blum (1956) for example notes that *Closterium acerosum* may occur on silt banks which, however, are often rapidly covered during times of high turbidity by new silt layers. Obviously this desmid's motility is of great advantage in this habitat (see p. 67).

Current studies of the algae of an English chalk stream, being conducted in the author's laboratory, have revealed that amongst large populations of attached, and some motile, diatoms associated with aquatic macrophytes, the only moderately common desmids are *Closterium acerosum, C. moniliferum* and *C. ehrenbergii*—again all motile species (Fotheringham, personal communication).

Desmids often form a component, though again a small one, of the phytoplankton community of rivers, especially in some of the world's larger, slow-flowing waters such as the Nile (Brook 1954, identified 24 planktonic species in the Blue and White Niles; see also Talling 1976), the Mississippi and some of its tributaries (Brook, personal observations) and especially the Amazon.

Many of the desmids found in the plankton of rivers are, however, chance or tychoplankters which probably originate in the static, or slowly flowing, regions (Brook and Rzŏska 1954, Schmidt 1970). These are probably supplied to the main stream somewhat erratically, depending on the prevailing hydrological regime. Where humid swamps bordering rivers are drained into them, as in the upper reaches of the White Nile, the Congo (Van Oye 1926) and the Amazon (Schmidt and Uherkovich 1973), desmids are often quite numerous. A great diversity of species has been identified from different regions of the Amazon. Förster (1969) recorded 409 taxa, while more than twice that number, in fact 966, have been listed by Thomasson (1971) (see also Förster 1974). In the Amazon itself, with its sometimes distinctly acid waters (pH 5.0–7.5), Schmidt and Uherkovich (1973) have found, in different regions of this great river system, that although there are often more species of desmids present in the phytoplankton than species belonging to other algal groups, this number ranges from 6 to 35 depending on the season and locality (see also Schmidt 1970).

It should be noted that in the transition from the lotic to the lentic state typical lake plankton appears to be rapidly filtered out, especially if weedbeds occur in the headwaters of the river. As found by Chandler (1937), there is a remarkably specific selection of forms and desmids are amongst the first group of the lake phytoplanktons to be eliminated.

In the River Thames, England, the plankton of which has been monitored weekly at Medenham, Oxfordshire by the Water Research Centre since 1965 (15 years), there have been only 29 occasions on which desmids have been observed in plankton counts. On 15 occasions the desmids were *Closterium acutum*, on 10 occasions an unidentified *Cosmarium* and a *Staurastrum* sp. only 3 times. Their abundance was never more than about 10 cells per ml (Lack, personal communication, also Lack 1971).

From the considerably polluted Susquehana River near Binghamton, New York, Wager and Schumacher (1970) list 9 species of *Closterium* from the plankton including *C. acerosum*, *C. moniliferum*, *C. striolatum* and *C. lineatum*, plus 4 *Cosmarium* and 3 *Staurastrum* species. In the upper Rhine and Neckar, Backhaus and Kemball (1978) have found 16 *Closterium* species and only 3 *Cosmarium* and one *Staurastrum* (*S. tetraedron*) in the plankton; but desmids have never been found to form a significant component of the total algal biomass.

DROUGHT RESISTANCE IN DESMIDS

Although the overriding majority of desmids are strictly aquatic and unable to withstand even moderate desiccation some, especially saccoderms of the Mesotaeniaceae, are sub-aerial and occur amongst bryophytes, on soil and peat and on dripping rock surfaces (West and Fritsch 1931, Grönblad 1932). Lund (1942) records *Closterium striolatum* and *C. malvinacearum* on the exposed marginal deposits of a small, somewhat alkaline pond (pH 7.0–8.4). The only detailed study to investigate the ability of desmids to withstand drying was that of Evans (1958, 1959). He examined the occurrence of the marginal algae of small ponds and found several desmids that were able to survive exposure and drying in the vegetative state. Thus the saccoderm *Netrium oblongum* var. *cylindricum* was frequently found on the litter of one pond some 2 m away from, and 30 cm above, the water's edge, where, after a period of exposure, the litter's moisture content had fallen to 55%. Species of *Cosmarium* and *Closterium* were found to have varying degrees of ability to survive experimental drying mostly by the formation of zygospores, though some species were resistant in the vegetative condition. Thus direct observation of *Cosmarium cucurbitum* after 54 days of drying, by which time the moisture content of the litter was 4.5%, was found to possess somewhat discoloured chloroplasts, prominent oil globules and a thin mucilaginous sheath. These drought resistant cells are considered by Evans to be functionally comparable with the zygospores of other members of the Conjugatophyceae. Although *Cosmarium praemorsum* and *C. botrytis* also survived prolonged and severe drying, Evans had no observations to indicate how they did so.

THE DISPERSAL OF DESMIDS

That some desmids are able to withstand quite considerable desiccation for

extended periods clearly suggests that at least some species may be successfully dispersed by wind and by animals that move from one body of water
to another, such as aquatic insects and waterfowl. There is a certain amount
of evidence that this indeed does occur. Thus Brown, Larson and Bold
(1964) found viable cells of *Cosmarium*, *Cylindrocystis* and *Roya* species
from atmospheric samples collected in Texas, U.S.A.

Evidence that insects may be involved comes from Iréné-Marie (1938),
who examined a number of dytiscid beetles in the Montreal region of
Canada and found specimens of *Closterium* in their claws; Parsons,
Schlichting and Stewart (1966) found *Cosmarium* raised from cultures
from dragonflies and damselflies; and Schlichting and Milliger (1969)
have demonstrated the possible dispersal of *Cylindrocystis* and *Penium*
by a hemipteran.

The first indications that birds could be agents of desmid dispersal
were again from observations made by Iréné-Marie (1938). He described
how, by washing the feet of a Blue Heron in filtered water, after observing
its flight between peat bogs, he counted a total of 126 cells of 15 different
species of desmids. He also killed ducks before they could alight in water
of peat bogs. Plumage washings from one duck yielded 517 desmids
representing 31 species of 31 different genera.

Proctor (1959) points out that transport on the feet, feathers or bills
of birds would not be very effective for dispersing desmids, most of which
do not survive even modest desiccation, over long distances. He suggests
that these algae might, however, be transported for considerable distances
and avoid desiccation, if they could survive the passage through the ali-
mentary tracts of migratory birds. Over a period of a year Proctor examined portions of the gut from 25 different migratory birds from W. Texas
and south central Oaklahoma. Viable algal cells were present in the lower
digestive tracts or caeca of one or more birds from each of the 25 genera
examined. Piscivorous birds (grebes, herons, kingfishers and egrets)
contained only simple '*Chlorella*-like' greens, but a considerable variety
of genera were recorded from ducks and bottom-feeding shore birds. As
many as 4 genera of desmids (*Closterium*, *Penium*, *Cosmarium* and *Staurastrum*) were recorded from the guts of some of the birds. From this study
Proctor concludes that many freshwater algae (including desmids) may
easily be carried between bodies of water 100–150 miles apart.

In a later paper, Proctor (1966) makes the point that neither his study,
nor those of Iréné-Marie (1938), provide experimental evidence to show
whether desmids attached to external surfaces can survive longer and
consequently can be carried greater distances than those taken into the
digestive tract. He draws attention to the suggestion of Loffler (1964)
that there may be aquatic organisms which are better able to withstand a

period in the digestive tract of birds than desiccation. To test this possibility, vegetative cells of 8 different species of desmids were fed to waterbirds—the common mallard duck and the killdeer. Viable cells of 2 *Cosmarium* spp., 2 *Micrasterias* spp., and a *Euastrum*, *Pleurotaenium* and *Staurastrum* species were recovered after being in the digestive tracts of both duck species for at least one hour. The only non-survivor was *Closterium parvulum*, though its zygospores seemed to pass unharmed through the tracts of 9 out of 11 killdeer. Since vegetative cells of all the desmids tested were readily killed by desiccation, Proctor suggests that the cells of most desmids can probably be carried further in waterbird digestive tracts than on external surfaces.

As mentioned earlier (p. 201), Heimans (1969) proposed the term 'accessibility' as a factor of some significance in accounting for the distribution of desmids. Thus he comments that it is often difficult when comparing species lists from apparently similar water types to produce a clearly defined group of characteristic species which typify that particular milieu. This failure in some cases, he states, can be because otherwise identical environments in separated localities differ considerably in their accessibility, with the result that many species which could survive and reproduce in them, in fact, never reach them. Heimans argues that 'in this case the environments in these different localities are not really identical because they differ in their "accessibility factors"'. He stresses the importance of the wefts of filamentous algae or of *Sphagnum* and *Utricularia* in desmid dispersal; also that bodies of water lying on the fixed migration routes of waterbirds are favourably placed by reason of their accessibility. This he believes accounts for the fact that recently formed ponds in the coastal dunes of the Netherlands, sometimes within ten years of their natural or artificial origins, contain numerous desmids that do not occur further inland but which are common in the distant north and north-east of Scandinavia (see also Tomaszewicz 1974).

DESMID DISTRIBUTION AND TEMPERATURE

Desmids occur over a wide range of water temperatures, diverse and abundant collections having been recorded, on the one hand, from tropical lakes, ponds and marshes, which may be well in excess of 30°C, to arctic regions on the other. Despite the fact that the water in many arctic ponds (e.g. the thermokarst ponds of Hobbie 1973), especially those of less than 2 m in depth, freeze solid during the protracted high latitude winters, their desmid floras are likewise rich and varied (Croasdale 1955, 1956, 1957, 1962, 1965, 1973). Hilliard (1966) has observed that desmids contributed to 42% of all species of algae recorded from an Alaskan tundra pond. Less abundant and considerably less diverse desmid floras would seem to occur in the Antarctic (Corte 1962, Fritsch 1912, Heywood 1977, Hirano 1965).

In these very cold regions it is quite obvious that many desmid species can withstand extended periods of freezing. Experiments by Duthie (1964), prompted by observations on desmid-rich sediments which were frozen solid for a period in excess of 70 days, revealed that, provided the cells are acclimatized to low temperatures (4°C) prior to freezing, their survival rate is high. A recent study of the metaphyton of a tundra pond in the N.W. Territories of Canada (Sheath and Hellebust 1978) indicates that three of the five most abundant species were the desmids *Cosmarium botrytis* var. *tumidum*, *C. reniforme* and *C. subcrenatum*. These showed a bimodal periodicity during the 101-day ice-free period, with one peak at the end of June, the second in mid-August. Growth expressed as biomass increased from 5–20 mg m^{-2} for the first two of the above species and from about 1–15 mg m^{-2} in the case of *C. subcrenatum* in June, when for much of the month water temperature was less than 8°C, and the maximum for a brief period was only 12°C.

Saccoderm desmids have been found, often in great abundance, as cryobionts on snow and especially the ice of glaciers in Alaska (Kol 1942) and other parts of the world (Berggren 1871, Borge 1899, Lagerheim 1892). *Ancylonema nordenskioldii* and *Mesotaenium berggrenii* seem to occur only on the ice of glaciers and indeed at times are in such profusion as to cause characteristically coloured purplish to brown 'ice blooms'. The sexual reproduction of both has been observed in this environment (Kol 1942, and see p. 238). *Cylindrocystis brebissonii* f. *cryophila* occurs not only on glacial ice but also on snow, having been noted as a permanent element of the cryophyte vegetation of Greenland, Alaska, Scandinavia, Siberia and Switzerland.

At the other extreme of the temperature range only one desmid would appear to have been found as a component of the flora of hot springs (Kullberg 1971). Thus *Cosmarium obtusatum* has been recorded at 38°C down a temperature gradient of two such springs in Montana, U.S.A. Numerous species of *Cosmarium* and *Staurastrum* appeared to be quite healthy in small temporary pools adjacent to the Blue Nile in the Sudan, in which temperatures of up to 40°C were recorded (personal observation).

THE GEOGRAPHICAL DISTRIBUTION OF DESMIDS

Because many desmid species seem to be remarkably sensitive to environmental conditions, especially water chemistry, and because they are not readily distributed from one body of water to another (see above), there are many which would appear to have a fairly restricted distribution.

On the other hand there are many taxa whose distribution can be said to be almost ubiquitous, e.g. *Closterium ehrenbergii, Cosmarium botrytis, Staurastrum pingue*. Although attention was drawn by the Wests (West and West 1904) to the fact that there would seem to be clearly American, African, Indo-Malaysian etc. desmid species, little true progress would seem to have been made in this field of desmid biology. Indeed, one can even now repeat the statement made by Prescott in 1952 in his review article on desmid biology 'the picture is still hazy and lines are not well drawn, partly because the literature is so bulky that summarizing analyses are difficult, whereas many species seem to be characteristic of geographical areas and indeed may be classed as endemic, generalizations are continually being broken down as information increases and as supposed endemics are reported from far away stations'. Moreover, it would seem that real progress as to the true geographical limits of taxa will not be achieved until more knowledge is available about the morphological variability, under a range of environmental conditions, of many desmid species and species groups. It might, however, be rewarding to pursue this field of research for as Fritsch (1935) states 'it is possible that a sound knowledge of the distribution of Desmids would shed considerable light on the question of the land connections existing in recent epochs of the earth's history'.

DESMID FLORAS AND LAKE TYPES

The factors that would appear to determine what will be the species composition and the eventual relative abundance of the components of the algal flora of a particular body of water are numerous and diverse and their interactions complex. Clearly, they must first reach the water, and this event will depend on its proximity to other algal (in this special case, desmid) habitats. Secondly, much will depend on the ability of particular species to survive the mechanism of transport by which they are dispersed (p. 207). Thirdly, having reached the body of water in a viable state, the final condition to be met is that the complex of chemical conditions, inorganic and no doubt organic, and physical conditions are appropriate not only for their survival but their growth and establishment as one component of a flora of interacting and competing populations, consisting of their own kind (desmids) and a whole range of other algae.

The British algologists W. and G. S. West (1909), as a result of extensive studies of the British freshwater plankton, were the first to recognise that desmids seemed to be associated with particular types of water. Indeed, they concluded that plankton populations containing the greatest diversity of desmid species correspond geographically with pre-Carboniferous rocks and occur in water that is particularly low in dissolved minerals. Although the Wests did not formulate definite biological lake types, it was partly on the basis of their results, and those of Lemmermann (1904) and Wesenberg-Lund (1905), that Teiling (1916) applied the terms Caledonian and Baltic, respectively, to lakes poor and rich in nutrients. Naumann (1917, 1919) later substituted the ecological nomenclature, oligotrophic (= poor feeding) and eutrophic (= good feeding) to poor and rich waters. These are far from being precise terms, but in relation to water chemistry the most oligotrophic lakes tend to be synonomous with very 'soft' waters, with little or no bicarbonate alkalinity, while at the other end of the trophic scale the more eutropic, enriched lakes are usually the richest in calcium bicarbonate. More recent usage based on measurements of primary production equates oligotrophy with low productivity, though it should be pointed out that low phytoplankton production may also occur in extremely hard waters with $CaCO_3$ supersaturation (cf. the Durness Lochs of Sutherland, Scotland, where most production is of benthic charophytes, Spence 1964, 1967). On the other hand soft waters can become highly eutrophic in this sense when enriched with nitrogen, phosphorus or potassium (Brook and Holden 1957, Brook 1958, Holden 1958).

Pearsall (1921, 1924, 1930, 1932) introduced the concept of lake evolution, and his chemical investigations of the English Lakes provided some indications as to factors that might be responsible for the appearance of distinctive phytoplankton types. Pearsall demonstrated a progression from oligotrophic (primitive) lakes, with a plankton dominated by desmids, through an intermediate stage, in which diatoms and desmids were most numerous, to eutrophic (silted) lakes dominated by diatoms and eventually blue-green algae.

The relative dominance of the major taxonomic groups of planktonic algae recognised by these earlier investigators is a phenomenon which has been used to indicate the trophic status of lakes (Thunmark 1945, Nygaard 1949). Although these have been found to be 'eloquently expressive' in providing reliable indications of the productive potential of bodies of water in Sweden and Denmark by Thunmark and Nygaard, respectively and also of British lakes (Brook 1965) and lakes in Minnesota, U.S.A. (Brook 1971), great care must be exercised in using this technique. Its major limitations have been described by Brook (1959a, 1965; see also Geelen 1969 quoted in Coesel 1975).

In Brook's investigations (1959, 1965, 1971) particular attention was paid to planktonic desmids and the trophic status of the lakes in which they occur. Satisfied that quotients, and in particular Nygaard's Compound Quotient, can be applied with reliability, provided the limitations outlined in his 1965 paper were observed, quotient determinations

Table 10. Range of alkalinity and compound phytoplankton quotients of lake waters in which planktonic desmids occur (Brook 1965).

Species	Compound quotient		Alkalinity range (ppm CaCO$_3$)
	Average	Range	
Staurastrum gracile	6.25	2.3–12.0	46.0–195
S. chaetoceras	6.0	0.9–20.0	4.0–194
S. planctonicum	3.2	0.7– 9.5	2.0– 78
S. pingue	2.6	0.4–11.5	7.0–200
S. furcigerum	1.8	0.4– 5.5	3.0– 78
S. cingulum var. *obesum*	1.2	0.2– 9.5	2.0– 42
S. arctiscon	1.2	0.5– 2.0	
S. lunatum var. *planctonicum*	0.96	0.2– 4.0	1.5– 38
S. pseudopelagicum	0.87	0.2– 2.3	1.6– 45
S. boreale	0.85	0.2– 1.5	
S. cingulum	0.83	0.3– 3.0	7.0– 89
S. denticulatum	0.79	0.2– 2.0	4.0– 38
S. anatinum	0.75	0.2– 2.0	1.0– 38
S. longispinum	0.71	0.2– 1.1	1.0– 12
S. brasiliense	0.70	0.3– 1.1	4.0– 8.4
S. longipes	0.59	0.2– 2.0	8.0– 45
S. ophiura	0.55	0.2– 0.8	3.0– 38
Staurodesmus dejectus	1.0	0.2– 3.0	4.0– 45
S. brevispinus	0.9	0.2– 3.5	1.0– 38
S. cuspidatus	0.85	0.2– 3.4	4.0– 78
S. curvatus	0.7	0.2– 2.0	1.8– 41
S. glabrus	0.7	0.5– 1.1	4.0– 9.4
S. megacanthus var. *scoticus*	0.67	0.2– 1.5	
S. jaculiferus	0.66	0.3– 1.1	1.8– 38
S. sellatus	0.64	0.2– 1.5	
S. megacanthus	0.6	0.2– 1.3	1.6– 16
S. subtriangularis	0.58	0.1– 1.3	1.6– 10
S. triangularis	0.53	0.2– 1.1	
S. aversus	0.4	0.2– 0.9	1.0– 9.0
Cosmarium granatum	4.8	1.1–11.5	12.0– 76
C. humile	3.4	0.5– 9.0	5.0–135
C. botrytis	2.75	0.7– 6.3	3.0–135
C. impressulum	2.1	1.0– 3.5	
C. depressum	1.8	0.5– 7.0	3.6–194
C. abbreviatum	0.95	0.2– 2.0	1.6– 26
C. contractum	0.67	0.2– 1.3	0.0– 26
Spondylosium planum	0.88	0.2– 2.3	3.0– 33
Cosmocladium saxonicum	0.71	0.2– 1.1	1.6– 16
Euastram verrucosum	1.0	0.2– 5.0	1.0– 20
Xanthidium subhastiferum	0.86	0.5– 1.1	5.0– 38
X. antilopaeum	0.76	0.2– 2.0	0.0– 38
X. controversum	0.61	0.2– 1.0	1.6– 16
Closterium ceratium	5.75	5.0– 6.5	114.0–194
C. aciculare var. *subpronum*	3.5	0.8–10.0	12.0–135
C. setaceum	0.79	0.3– 2.0	1.0– 45
Micrasterias sol	0.7	0.2– 5.0	1.0– 20

were made for some 300 lakes of widely differing types in various parts of
the British Isles. From these the average compound quotient has been
calculated for each of the most commonly occurring desmid species, along
with the range of trophic levels over which they have been found and the
alkalinity range expressed in ppm $CaCO_3$ (Table 10).

A similar analysis was conducted for 200 Minnesota lakes many of
which are much more eutrophic than British lakes having suffered from
very considerable artificial enrichment ('cultural eutrophication'). Four
distinct groups of lakes, largely reflected by their salinities, can be recognised (Bright 1968). Those termed Group I lakes occur on non-calcareous
drift and have salinities between 10–50 mg/l; those of Group II are
on calcareous drift with salinities from 70–200 mg/l; the Group III are
highly saline, yielding a total anionic content of 400–700 mg/l. The lakes of
Group I and II are bicarbonate waters, while in marked contrast, those of
Group III are exceptionally rich in sulphates. The majority of the lakes
containing desmids in their plankton, of which 61 taxa were identified, were
found in the Group I lakes. A few of the lakes contained over 20 limno-
plankters. In contrast most of the Group III and many of the Group II
lakes contained only one or two species; a few had more. Four species,
Staurastrum anatinum fac. *denticulatum-paradoxum*, *S. pingue*, *S. planctonicum* and *Staurodesmus mamillatus* were found in nearly 50% of all
the lakes investigated. As in the studies of British lakes (Brook 1965),
several desmid species clearly had eutrophic tendencies and *Closterium
ceratium*, *Cosmarium depressum*, *Staurastrum cingulum* (Windermere
type) (see Brook 1959) and *S. cingulum* var. *obesum* occurred only in Group
II lakes. Presumably tolerant of high salinities, and often considerable
entrophication, since their distribution extended into the Group III lakes,
were *Closterium acutum*, *C. aciculare* var. *subpronum* and also *Staurastrum
chaetoceras*, *S. contortum* and *S. excavatum*. As can be seen in Fig. 74, in
which the range and average compound phytoplankton quotients are
indicated, although they differ slightly, it is very reassuring that the ordering of the trophic preferences of all the species common to Britain and
Minnesota is the same. Also it is very interesting to note that the average
compound quotients of the Minnesota desmids are consistently higher
than the same species in British lakes.

An analysis of the trophic preferences of desmids occurring in the
British lake plankton has shown that, although confirming the generally
accepted conclusion that the greatest number of desmid species (59%)
most frequently occur in oligotrophic lakes, nearly a quarter (24%) of
British forms are most frequently associated with distinctly eutrophic
waters (Brook 1965, Table 7). This finding points to the fact that the inclu-
sion of all desmid species present in a sample when determining compound

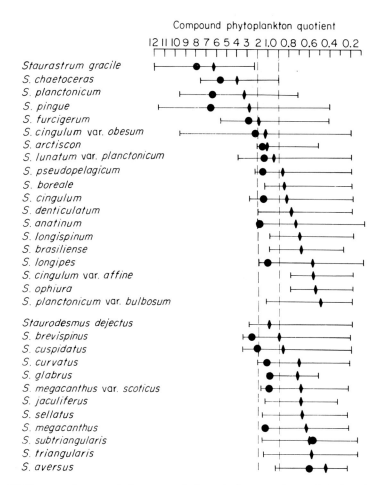

Fig. 74. Range and average (♦) compound phytoplankton quotients of planktonic desmids occurring in British lakes. The circles (●) show average compound quotients for desmids in Minnesota, U.S.A. lakes.

quotients decreases the reliability of this method of assessing trophic status. Indeed, it adds weight to the view (Brook 1959b) that a quotient should be formulated based only on a limited number of planktonic desmids (and other algae) whose status in the plankton and especially whose nutritional requirements have been adequately investigated.

DESMIDS IN LAKE SEDIMENTS AND LAKE HISTORY

In attempts to reconstruct the changes which are presumed to have

P

occurred in post-glacial environments special attention has been directed towards the analysis of the occurrence of diatom frustules in lake sediments. There has, however, been a recent attempt to relate desmids to lake history by Frederick (1977) who has studied these and other algae in the sediments from 2 bogs, and a lake, in central Ohio. As he points out, Borge (1892, 1895, 1896) was the first to investigate in any detail the occurrence of desmids in post-glacial sediments, and these studies were followed by the examination of other European sediments by Weber (1918) and Messikommer (1938), and more recently of Russian lake sediments by Korde (1960, 1966). Korde has indicated that diverse algal remains, other than diatoms, can be remarkably well preserved and that the study of the occurrence of particular algal groups (Chlorococcales, Desmidiaceae etc.) rather than individual taxa, is sometimes a more valuable approach for determining the environmental conditions of lakes over a period of time. It might be presumed that the differential preservation of organisms in sediments may cause difficulties in the interpretation of the results of sediment core analyses, but results suggest that an adequately representative proportion of those algal groups commonly used as environmental indicators are sufficiently preserved to allow for the reconstruction of the history of many lakes. Frederick (1977) has applied the methods used to determine the trophic status of extant ponds and lakes (compound quotients and diversity indices) to the other-than-diatom algal floras of post-glacial sediments. Although he has failed to restrict the determination of his quotients to phytoplankers (see limitations pointed out by Brook 1959), he would appear to have demonstrated that it is possible to follow algal community succession in sediments throughout their post-glacial history and that the occurrence of the desmids in these can be particularly instructive.

DESMIDS AND WATER CHEMISTRY

The only significant investigations carried out to date to explore the responses of desmids to the possible chemical conditions which may determine their distribution in waters of different trophic status are those of Moss (1972, 1973a, b, c). In these studies he examined the growth responses in culture of desmids associated with both eutrophic and oligotrophic waters. As species typically occurring in the former he used *Mesotaenium kramstai*, *Closterium acerosum* and *Cosmarium botrytis*; as oligotrophic desmids he selected *Desmidium swartzii*, *Pleurotaenium trabecula*, *Gonatozygon monotaenium*, *Micrasterias americana* and a *Roya* sp. He looked first at the influence of calcium levels and the ratios of monovalent to divalent cations as possible controlling factors, for both have become almost a part of phycological dogma as being especially significant in determining the distribution, especially of oligotrophic desmids.

The first results of his growth–response experiments would seem to cast considerable doubt on the importance of mono- to divalent cation ratios, since over a range of 12.2–0.110, their effect was of no significance. Indeed, oligotrophic desmids were unaffected by Ca and Mg levels comparable with those of very hard water lakes. An examination of the effects of increasing Ca levels alone was explored with *Gonatozygon, Roya, Desmidium* and *Micrasterias*, all oligotrophic species. All grew well at Ca levels between 10 and 100 mg Ca^{2+}/l and indeed at very low concentrations (0.02 mg/l) the growth of *M. thomasiana* was reduced by approximately a half to about one division in 10 days. Moss also reports that 0.1 mg Ca^{2+}/l appeared to be an adequate concentration for all the desmids tested, except *Desmidium swartzii* for which even 1.0 mg Ca^{2+}/l was insufficient; also the growth rate of *Roya* was significantly reduced at 1 mg Ca^{2+}/l. Indeed, the species requiring the highest minimum Ca levels were filamentous oligotrophic species, and Moss suggests that their relatively high requirements of between 1–3 mg Ca^{2+}/l would be satisfied in most waters; also that it may relate to the filamentous habit of these desmids.

Tassigny (1971) has also studied the growth of desmids in relation to the Ca^{2+} levels in culture and found that *Staurastrum paradoxum* (?) was indifferent to Ca levels, while *S. sebaldi* var. *ornatum* increased its doubling time from 6 days at 10.3 mg Ca^{2+}/l to 7.25 days at 20 mg Ca^{2+}/l. There was a similar modest decline in growth rate of *Closterium strigosum* from 2.33 days in 2.7 mg Ca^{2+}/l to 3.3 days in 54.8 mg Ca^{2+}/l, and with *Micrasterias crux melitensis*, a species more oligotrophic than the others investigated, its growth rate fell from 4.5 days in 10.3 mg Ca^{2+}/l to 6.0 days in 20 mg Ca^{2+}/l.

In a study of the nutrition of *Cosmarium turpinii* Korn (1969) determined the optimum concentrations of the elements Ca, N, P, S and Mg for growth, and for this desmid Ca would appear to be the most critical element, with a distinct growth peak at the optimal concentration. Because Korn observed a very rapid bleaching of the *C. turpinii* cells in Ca-deficient conditions this he interpreted as a consequence of ion imbalance leading to cell-membrane denaturation. The desmid's fairly high Ca requirement would seem to place it in the category of a hard-water desmid growing best in somewhat alkaline conditions (pH 7.5 optimum). Korn also found that Fe, Bo and Mn were essential micronutrient elements, but that Zn, Cu, Mb and Co had an inhibitory effect on the growth of *C. turpinii*.

Gough (1977) reports from experiments on the growth of some desmids at different levels of calcium and pH that *Triploceras gracile* isolated from an acid bog, hence presumably a species with oligotrophic tendencies, grew better at Ca^{2+} levels of 3 mg/l than at 50 mg/l, and at a pH of 6.0 in preference to 8.5. In contrast *Closterium moniliferum*, isolated from *Myriophyllum spicatum* growing in a hard-water lake, grew equally well at the two Ca levels but its growth was better at the higher pH level (see p. 218). The growth of *Cosmarium granatum* from the same habitat was better at both the higher pH and Ca concentration.

In a study of desmid distribution in a wide range of Estonian lakes, Kovask (1973) states that only *Closterium navicula*, *Cosmarium ornatum* and *Tetmemorus granulatus* occur in concentrations of 10 mg/l Ca^{2+}. In agreement with Brook's (1965) study, he found that a considerable number of desmids at high Ca concentrations (40–60 mg/l) and his list of such taxa contain many species in common with Brook (see also Messikommer 1928, 1942). Indeed, Kovask (1973) questions whether Ca is a limiting factor in desmid distribution and his results suggest that 'The total mineral content' which, however, he expresses as HCO_3 mg/l probably exerts a more significant controlling influence (see p. 219 and Fig. 75 below).

In summary, there would seem to be no clear evidence that desmids are calciphobic, though high levels of calcium may adversely affect the growth of some oligotrophic species. Lund (1965), in his review of phytoplankton ecology, remarks that algae have often been distinguished as calciphilous and calciphobic but adds that there would seem to be relatively few planktonic forms whose distribution is restricted by amounts of calcium. He adds that 'in some of the examples commonly quoted, such as desmids, other factors may be equally important'. These other factors are discussed below (p. 219).

Lund (1965, p. 254) quotes a Russian study (Braginskii 1961) in which K salts were added to fish ponds (also N and P). Potassium alone did not produce any increase in rates of primary production over controls, but desmids became predominant in place of blue-greens or coccoid greens. The reason for the observed effect of K was uncertain; as Moss (1973a) comments, 'These experiments appeared inadequately controlled to establish unequivocally the influence of K.' Indeed, no case can yet be made that major cation levels are involved in producing the qualitative differences that unquestionably do exist, expecially between the desmid floras of oligotrophic and eutrophic waters.

The nitrogen metabolism of desmids has received little attention and as Villeret (1973) points out the utilization of nitrate presupposes the intervention of nitrate reductase, the presence of which has been demonstrated (Villeret and Savouse 1962, Van der Ben 1970). The possibility that the N of uric acid and glyoxylic urides may be utilized has also been shown for some desmids.

A study by Nakarishi and Monsi (1976) of some of the factors which may control the species composition of freshwater phytoplankton suggested that two groups of algae were distinguishable with respect to inorganic-N levels. Those occurring in, and isolated for study from, eutrophic lakes showed significant decreases in chlorophyll 'a' as nitrate-N levels were reduced. At less than 0.1 mg/l the pigment level was such that photosynthetic activity more or less stopped. Algae from an oligotrophic lake, which included a *Cosmarium* sp., maintained the amount of chlorophyll, and hence their photosynthetic activity, even in media where the nitrate-N level was well below 0.1 mg/l. It would be of great interest to explore how eutrophic desmids respond to reduced N levels.

THE PH–FREE CO₂–BICARBONATE SYSTEM AND DESMIDS

It has been recognized for some considerable time that some aquatic plants can utilize only free CO_2 as a carbon source for photosynthesis while others can use bicarbonate (Ruttner 1953, Raven 1970). Shapiro (1973) has provided substantial evidence that the role of free CO_2, and its effect on pH and, in turn, the availability of bicarbonate, may relate to the success, or otherwise, of blue-green algae which commonly produce extensive water blooms in eutrophic lakes. Thus, he demonstrated in a series of experiments using large plastic bags in lakes, that natural populations of cyanophytes could be replaced by chlorophycean algae when, even after enrichment of the enclosures with phosphorus and nitrogen, CO_2 was added, or pH reduced. This addition altered the CO_2–bicarbonate equilibrium, thus making CO_2 available for photosynthesis especially by Chlorococcales, increasing their rate of growth so that they out competed the blue-greens. The addition of phosphorus or nitrogen alone increased the growth of the blue-greens preferentially.

Moss (1973a) conducted a series of experiments to explore the effects of changes in the pH–CO_2–bicarbonate system on desmids (and other freshwater algae). One of his findings was that distinct differences seemed to exist in the maximum pH's tolerated by eutrophic, as distinct from, oligotrophic algae. However, all the desmids investigated grow at pH values well above neutral, so that the maximum growth of *Micrasterias denticulata* and *M. thomasiana* lay between pH 7.65–8.1 and 7.7–7.75 respectively. The eutrophic *Mesotaenium kramstai* and *Closterium acerosum* grew successfully at levels above 9.0. At the acid end of the pH scale no particular pattern was established. In summarizing his results Moss states that, directly or indirectly, pH seems likely to be an important factor in determining why oligotrophic species do not grow in hard-water lakes. Like Shapiro (1973) he stresses that the most likely explanation lies in the

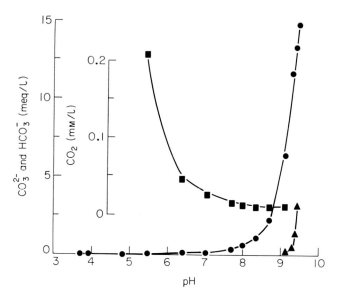

Fig. 75. Amount of free CO_2 (■), bicarbonate (●) and carbonate (▲) in relation to pH in experimental media used in desmid growth studies (from Moss 1972b).

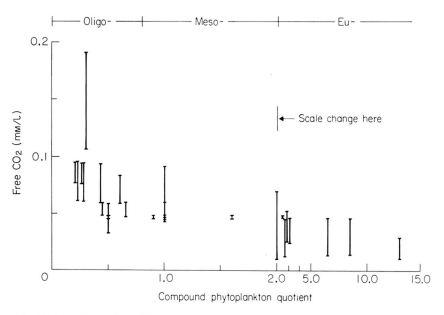

Fig. 76. The relationship of free CO_2 levels to the compound phytoplankton quotient in 26 Scottish lochs from data of Brook (1960) (from Moss 1972b).

different availabilities for photosynthesis of inorganic carbon sources (free CO_2 or bicarbonate) at different pH values. Moss thus found no growth of those desmids considered to be oligotrophic, or soft-water, forms above pH 8.6. As indicated in Fig. 75 this corresponds to the hydrogen ion concentration at which all free CO_2 disappears. Indeed, most growth of *M. denticulata* and *M. thomasiana* ceased when free CO_2 was about 58 % higher than the minimum, and *Gonatozygon monotaenium* grew much less successfully at pH 8.85 than at pH 8.4, between which values there is a difference in free CO_2 of about 11 %. The observed difference in tolerance of the *Micrasterias* cells compared to those of *Gonatozygon* is ascribed by Moss to the bulkiness of those of the former. He proposes that higher external levels of free CO_2 might be necessary to provide for its sufficiently fast diffusion into the chloroplasts—an interesting speculation on the significance of cell morphology in relation to ecological success. That there are desmids that will tolerate high pH levels, and so presumably must use HCO_3 for photosynthesis, was demonstrated by experiments with *Closterium acerosum*. As with other eutrophic algae this desmid would not grow at pH 9.2 or above, and Moss suggests that at this level carbonate toxicity may have then limited its growth (see Fig. 75). Experiments carried out in 1977 by Freshwater Biological Association phycologists on Lund-Tubes in Blelham Tarn (Lack and Lund 1974) produced (after enrichment with phosphorous, nitrogen, silica and iron) not the anticipated growths of blue-green algae but a community which in early September was dominated by the planktonic desmid *Staurastrum pingue*. This ubiquitous species, which is one of the few desmids to occur in very eutrophic lakes (Brook 1965, 1971), attained densities of up to 2000/ml at certain depths in the tubes (Reynolds, personal communication). The pH increased rapidly during the early and middle phases of *Staurastrum* growth increasing from ca 8.6 to 10.2. In a letter, Reynolds states that the pH 10.2 was close to the theoretical maximum based on alkalinity and says 'I would stress that at least the last division took place at this higher pH value and the population was maintained for 2–3 weeks at the same high pH level'. Full details of this fascinating experiment have been published by Reynolds and Butterwick (1979).

As Moss (1973a) points out, although oligotrophic species can grow at pH values typically above 8.0 in eutrophic waters if more than 0.011 meq/l free CO_2 is present, growth would be slow and these species would almost certainly be at a serious disadvantage in competition with eutrophic species. However, desmids considered to be oligotrophic do occur in eutrophic ponds and lakes (cf. Fig. 72). The possibility cannot of course be excluded that they are distinctive races with special tolerances but such a hypothetical explanation does not have to be evoked. Thus Moss (1973) found *Pleurotaenium trabecula* on highly organic sediments in Abbot's Pond (Somerset, England) and, although the pond is markedly alkaline and its pH is often greater than 8.0, he suggests that bacterial decomposition may lower the pH of the interstitial water to 6.0 or less, their respiration (fermentation?) increasing the free CO_2 level by a hundred-fold (Young and Cairn 1971). The possibility that such subtle micro-habitat differences occur would seem largely to have been ignored in investigations of desmid distribution.

Moss has reworked Brook's data (Brook 1965, Table 6) in which Nygaard's Compound Phytoplankton Quotient in relation to alkalinity is presented (see p. 213 and Fig. 76). In Fig. 76, reproduced from Moss (1973a), the highest and lowest calculated free CO_2 concentrations of each of 26 Scottish lochs have been compared with the quotients determined for each loch. A marked transition to low free CO_2 levels occurs associated with a quotient of just less than 1.0. This provides further support for the notion that there is a link between relatively high free CO_2 concentrations and oligotrophic desmids, and low free CO_2 levels and those desmids with meso- or eutrophic tendencies.

Moss (1973a) also considers the findings of Talling and Talling (1965) that in African waters 'an appreciable desmid plankton does not occur in waters with more than 2.5 meq/l of weak acid salts'. In Lake Malawi they found that desmids survived in pH levels which reach 8.5, the value approaching the upper extremes tolerated by oligotrophic desmids in Moss's experiments. The Tallings suggest that it was the relatively high ratio of mono- to divalent cations in African waters that allowed the desmids to tolerate somewhat higher levels of alkalinity (2.36–2.6 meq/l in Malawi); they feel that when Ca and possibly Mg is replaced by Na as the dominant cation, then desmids can grow in considerable alkalinities. Moss reasons that it is more likely that sufficient free CO_2 is available in these African waters at alkalinities up to 2.5 meq/l to allow the 'growth of species unable to use HCO_3, or free CO_2 at very low levels, for photosynthesis'.

Moss also draws on the example quoted by Hutchinson, Pickford and Schumann (1932) who found 30–40 species of desmids in a South African pan at pH 9.0 and alkalinity 1.3 meq/l. But, as Moss points out, these conditions were temporary and the high pH was clearly due to intense macrophyte photosynthesis. Extremely high pH levels were also experienced during nutrient enrichment experiments in oligotrophic Scottish hill lochs (Holden 1958) when blooms of the green alga *Dictyosphaerium pulchellum* caused mid-day pH values to rise to greater than 10.00 in the poorly buffered waters. A few desmids, however, still survived these conditions even though the normal pH of the lakes was less than 6.00 (Brook 1958) (see how p. 223 below).

Support for the importance of changing balance between CO_2–bicarbonate as a factor controlling the distribution of desmids comes from bioassay experiments performed by Moss (1973c). In these, selected test species of desmids were added to lake waters of different types. Both *Pleurotaenium trabecula* and *Micrasterias americana* grew much better in hard water when the pH was reduced and hence the equilibrium moved in favour of free CO_2; in a complementary experiment the growth of both desmids was reduced in soft waters whose pH had been increased. The observations of Brook and Holden in the series of fertilization experiments in Scottish hill lochs would also seem to be significant in this context.

In Loch Kindardochy, Perthshire, with an alkalinity of between 0.3 and meq/l, and pH values ranging between 6.8 and 7.8, although the addition of calcium superphosphate fertilizer caused the total phytoplankton population to increase by a factor of up to 8 times its prefertilization abundance, the specific composition of the loch's desmid flora remained unchanged. It is interesting to note in view of Reynolds' experience with the Lund Tubes in Blelham (see p. 221) that *Staurastrum pingue* increased from less than 10 to more than 50 cells per ml. In experiments in very oligotrophic lakes in Sutherland with prefertilization pH levels in some ranging from 5.5–6.0 (alkalinities—meq/l) and less oligotrophic ones from 7.0–7.5 (alkalinities—meq/l) pronounced blooms of Chlorococcales and some Cyanophyceae developed and their intense photosynthetic activity caused pH in two of the lochs to rise at times to as high as 10. The Compound Phytoplankton Quotients of these lochs, both treated with an N P K fertilizer, increased from levels clearly indicative of oligotrophy (0.3 and 0.7) to values suggesting considerable eutrophy (6.0 and 11.0 respectively). While the change in quotient was in part due to species which contribute to the quotient's denominator, it was also in part due to the disappearance of desmids from the plankton; on Moss's interpretation because of the non-availability of free CO_2 as a consequence of the photosynthetically induced pH change (Brook 1958, 1965).

These latter observations suggest that desmids are particularly good indicators of water quality, and evidence in support of this view comes from studies of the phytoplankton of Loch Leven, Scotland (Brook 1965) whose waters have been enriched by agricultural and urban development during the past 70 years. Net samples of phytoplankton collected by West and West in August 1904 during the Bathymetrical Survey of the Scottish loch (Murray and Pullar 1911) were found in the Royal Scottish Museum, Edinburgh. Seven species of planktonic desmids were present in the 1904 samples; the compound quotient was then $11/7 = 1.6$. In samples taken 50 years later in 1954 (and 1955) only four desmid species have been found while the number of species of Chlorococcales and Cyanophyceae has

increased considerably. The quotient determined for the 1954 and 1955
samples was 29/4 = 7/2. The desmid species recorded in 1904 and in
1954–55 were:

1904

Closterium aciculare
Cosmarium depressum var. *planctonicum*
Spondylosium planum
Staurastrum cingulum var. *obesum*
S. lunatum var. *planctonicium*
S. sebaldi var. *ornatum* f. *planctonicium*
S. pingue

1954–55

Cosmarium depressum var. *planctonicum*
Staurastrum cingulum forma
S. pingue
S. chaetoceras

It is significant in relation to the Moss–Shapiro hypothesis that Loch
Leven has increasingly produced large blooms of blue-green algae in
recent years. Almost 10 000 filaments/ml of *Oscillatoria bornetii* were
recorded in the summer of 1937 (Rosenberg 1938), 20 000 of *O. limnetica*
in the early summer of 1954, and 17 000 of *Aphanizomenon flos-aquae* in
1963 (for more recent data on the phytoplankton of Loch Leven see
Bailey-Watts 1974.) Such observations, and those of Coesel (1977),
suggest that desmids are sensitive indicators of environmental change (see
also Lund 1973, Hori and Ito 1959, and p. 227) and thus of value in
monitoring environmental change.

Evidence confirming the view that desmids respond dramatically to
changes in water chemistry is contained in the recent paper by Coesel
Kwakkestein and Verschoor (1978), which outlines changes that have
occurred in a complex of 26 shallow moorland lakes and ponds in Holland.
An extensive survey of the desmid flora of the lakes was carried out
between 1916–1925 by Heimans (1925) who recorded some 250 desmid
species. Even during the course of this early study Heimans detected a
gradual disappearance of some of 'the most exclusive species'. A survey
of the same area between 1950–1955 indicated that this tendency was still
occurring despite environmental conservation measures (Heimans 1960).
Part of the deterioration was caused by enrichment from domestic sewage
and although this pollution was halted and the saprobic mud deposits
covering the lake sediments were removed, these control measures during
the period of study produced no return of the previously rich desmid flora.
Coesel *et al.* (1978) examined the lakes and ponds again in 1975 and re-

ported that the desmid flora continued to be depleted. For example, the waters in the Central Vermen lakes and those of the Achterste Goorven contained 195 species in 1925, 123 in 1955, and in 1975 this was reduced to 68. However, as Coesel *et al.* point out, not only is the absolute decrease in numbers revealing, but of equal interest and significance is the fact that there has been a 'shift' in species composition. Thus considerable numbers of typically mesotrophic forms have disappeared and replaced by a smaller number of species with oligotrophic tendencies.

Part of the deterioration can clearly be attributed to eutrophication and is comparable to the changes in phytoplankton desmids in Loch Leven and elsewhere (see p. 223 above). However, the most surprising feature of the Dutch study is that 'oligotrophication', to quote Coesel *et al.*, is largely responsible for the impoverishment of the desmid flora. The change they ascribe in the main to a lowering of pH by 'acid rain' from industrialized areas consequent on high concentrations of sulphur dioxide and nitrous oxides in the atmosphere. A second factor, but which may be even more important than the acid rain in some areas, is the increasing effect of over shadowing by developing forests planted during the 19th century. Their influence is most marked in small pools and is not just the direct effect of shading on desmid growth but rather because the submerged macrophytes which have in consequence disappeared provide the most important substratum for the desmids of the metaphyton (cf. Bland and Brook, 1975 and p. 200). The probable effects of air pollution and re-afforestation on the desmid flora are shown in Fig. 77.

As pointed out by Hosiaisluoma (1976) airborne HCl pollution is increasing, and he has studied the effects of such pollution from a PVC factory on the vegetation of a raised bog in S. Finland. Here he found a mass occurrence of the saccoderm *Netrium digitus* in a pool slightly polluted by HCl. He used this desmid in experiments designed to examine the effects of a range of concentrations of Cl^- ions, added not only as HCl but also as NaCl, since, as he points out, the latter originating as sea spray is one of the most important and abundant compounds carried by wind and rain 'to ombrotrophic mire sites'. Such waters in S. Finland have a range of 1.8–12.4 mg Cl^-/l, with an average concentration of 5.4 mg/l. Hosiaisluoma lists the factors influencing the concentration of this ion in such areas. Of the cations present, Na^+ and Mg^+ originate from windborne sea spray, their concentrations generally increasing towards the coast. As indicated by Gorham (1956), the proportion of Cl^- ions of marine origin may be estimated approximately from the Na/Cl ratios which exist in natural waters (Na/Cl ratio in sea water $= 0.56$). In the Finnish bog pool the Na/Cl ratios averaged 0.40, suggesting that it was subject to HCl atmospheric pollution. As Hosiaisluoma points out, the

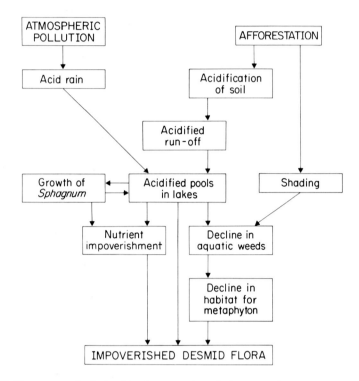

Fig. 77. The presumed relationships between air pollution and afforestation on the desmid flora of small lakes (after Coesel *et al.* 1978).

only ion to exceed the average value reported for other Finnish bogs was Cl⁻ which was present in concentrations of about 4.0 mg/l. Thus, if the Na/Cl ratio is assumed to be at least 0.5 in waters receiving Cl⁻ from the sea, about 3 mg Cl⁻/l must be of marine origin and 1 mg Cl/l from atmospheric HCl pollution. In this pool where the pH was 4.1 *Netrium digitus* grew abundantly along with *Zygogonium ericetorum* (Zyhnemaceae) yet in a nearby one where no *Netrium* was found, the Cl⁻ concentration was 4.6 mg/l and the pH 3.9. In a third pool in which there were a few *N. digitus* cells the pH was again 4.1, and the Cl⁻ concentration was 3.5 mg/l, 0.5 mg of which is considered to have originated from pollution. These observations seemed to indicate that the growth of this saccoderm was adversely affected by HCl, but not NaCl, a conclusion which Hosiaisluoma tested experimentally. HCl added to *Netrium* cultures in bog-pool water showed that the maximum Cl⁻ tolerance of the desmid was 10 mg/l (pH = 3.3–3.5). Cultures with an excess of 15 mg Cl⁻/l soon died. In experiments conducted in a nutrient solution, the desmid's growth and tolerance to

HCl was greater. Indeed, additions of 3–5 mg Cl^-/l as HCl promoted growth, for the reason that this resulted in producing what appears to be the optimum pH for *N. digitus* (pH 4.5–5.5). No divisions of the cells occurred when Cl^- concentration exceeded 12–13 mg/l, and the pH was hence below 3.5. At concentrations of greater than 10 mg Cl^-/l, the desmid cells often became greatly elongated but failed to divide, so that lengths of up to 800 μm (4× normal) were measured (cf. Moss 1973b, pp. 163 and 173).

The addition of NaCl to a natural water does not directly affect its acidity as does HCl, and Hosiaisluoma's culture studies showed that salt additions stimulated growth in bog-pool water significantly more than in nutrient medium. This he attributes to the more favourable pH (4.1–4.2) of the former than the latter (6.0–6.5), since *N. digitus* appears to be clearly an acidiphilous desmid. The optimal addition to the bog-water cultures was 20–30 mg Cl^-/l, compared with 15–20 mg/l in the mineral medium. Moreover, the desmid was found to continue to divide in concentrations of up to 700 mg Cl^-/l, and survive at 1000 mg Cl^-/l in the natural media, while in artificial media the upper concentration for growth was 350 mg/l, and for survival 400 mg/l. Other desmids, such as *Staurastrum pingue*, appear to be much more sensitive to NaCl (Soeder *et al.* 1971).

DESMID TOLERANCES TO HEAVY METALS

Of considerable interest in relation to desmids and pollution is a paper by Url (1955) on the unusually high resistance which desmids show, compared with phanerogams and to a lesser extent filamentous Zygnemaceae, to heavy metals. In this he reports on the effects, on a number of desmid species, of zinc, manganese, vanadium, chromium and copper when added to experimental cultures in the form of sulphates. Remarkably high resistance was found against Zn, Mn, V and Cr, though not against Cu, and the following species were found to be particularly tolerant: *Euastrum oblongum* and *E. verrucosum*, *Cosmarium pachydermum*, *Pleurotaenium truncatum* and *Desmidium schwartzii*. Many of these species survived the highest concentrations of the metal sulphate applied, which ranged from 0.00001% to as high in some experimental treatments, for example Mn, as 10 or even 20%. Much less resistant were the saccoderm *Netrium digitus* and *Micrasterias rotata*, *M. fimbriata* and *M. americana* and, in particular, *Closterium lunula*.

One of the most intriguing aspects of Url's research was the discovery of so-called 'death zones', especially in the case of chromium and vanadium resistance, and to a lesser extent with respect to manganese. Thus when *N. digitus* was subjected to a range of concentrations of $Cr_2(SO_4)_3$, the cells survived exposures from 0.0001 % to 0.001 % but from 0.005 to 0.1 % the cells died. However, at higher concentrations, from 0.5 to 5%, many of the cells survived. Similar results were obtained with *M. rotata, M. fimbriata, Staurodesmus convergens* and *C. lunula* in vanadium sulphate (VSO$_4$) and with *C. lunula* in MnSO$_4$. The latter's 'death zone' was, however, between 0.01 and 3.0% in this heavy metal salt.

'Death zones' have been discussed in some detail by Biebl and Rossi-Pilhofer (1954), and in attempting to explain this remarkable phenomenon they quote Kaho (1933). He is of the opinion that the heavy metal salts produced on the surface of the cell cytoplasm a type of protective 'skin' in the form of an irreversibly coagulated surface layer, preventing the entrance of the salts, at least for some considerable time. Hofler (1951) has suggested that in certain cases the various resistances of the cytoplasm of different species depends on the nature of their plasmalemmas, whose particular properties may not be present initially, but which develop, at least in certain cases, as a consequence of their being subjected to high heavy metal salt concentrations. This explanation would seem to be given support by Url's desmid observations as, for example, the case of *N. digitus* whose protoplast can tolerate concentrations of 0.0005–0.01 % Cr$_2$ (SO$_4$)$_3$ and in which a concentration of 0.05% is still too low to cause the development of the presumed coagulated protective layer in the plasmalemma. However, 0.1% would appear to be a critical concentration, while 0.5% would seem to be optimal for its formation, for above this concentration an increasing proportion of individuals die off, despite the plasmalemma being sealed. It is also possible that at the higher concentrations the chromium permeates the inner cytoplasm and kills it before the postulated protective layer can form.

In the case of those heavy-metal-resistant desmids which do not appear to have a 'death zone' their response is what has been described as an 'either-or' type: *either* there is no higher 'life zone' above a particular concentration, *or* a protective layer is formed which is totally impermeable, so that even marked plasmolysis of the cells, lasting for days, can be tolerated.

Url (loc. cit.) points to the interesting fact that, especially at the higher concentrations, the solutions of the trivalent metals chromium and vanadium produced a very acid reaction (Cr = pH 3.92 at 0.01% and 2.41 at 10%: V = 3.64 at 0.005% and 2.33 at 5%). Since these are the metals which in solution produced the definite 'death zones' it would seem that pH could be a crucial factor in so dramatically altering the permeability of the desmid cells outer membranes.

From an ecological point of view it would be of interest to explore the possibility of the existence of resistant desmid ecotypes, comparable to those known to exist in the case of higher plants, occurring for example, in streams and rivers in heavy metal-rich areas, such as North Wales and Teesdale in the U.K.

DISSOLVED ORGANIC SUBSTANCES AND DESMIDS

Tassigny (1971) and Moss (1973b) have examined the vitamin requirements of some desmids, and of those investigated none would seem to require biotin or thiamin. Both agree, however, that the oligotrophic *Gonatozygon monotaenium* and *Desmidium swartzii* had a B_{12} requirement. So also did the eutrophic species *Closterium acerosum*, *Micrasterias americana* and an oligotrophic *Cosmarium* sp. However, *C. botrytis* and *Mesotaenium kramstai* has no such requirement.

Korn (1969) would appear to have been the first to show a vitamin B_{12} requirement for desmids; in *Closterium turpinii*. Maximum growth stimulation was obtained with 5,6-dimethylbenzimizole nucleotide, and very good growth with other benzimizole derivatives. Their particular pattern of specificity, according to Korn, places *C. turpinii* into the *Escherichia coli* group of auxotrophic organisms (Droop 1962), a capacity unique (at least to date) for a member of the Chlorophyta. No absolute replacement for B_{12} was found, but cobalamin-coupled compounds were shown to enhance growth, so he suggests that these supplements may have important ecological implications in environments where B_{12} levels may be low.

There is some evidence that organic substances and especially those excreted by blue-green algae may be inhibitory to desmid growth (Lefèvre, Jakob and Nisbit 1900, Lefèvre 1964). However, many desmids seem to survive in ponds, lakes and cultures alongside large populations of certain Cyanophyceae. From what has been stated already about the effects which the latter, by their phytosynthetic activity, may exert on the balance between free CO_2 and bicarbonate (p. 219) it would seem that great discretion must be exercised in invoking the presence of these algae to account for the disappearance of desmids. In the study of the distribution of desmids in meromictic lakes (Bland and Brook 1974) where large metalimnetic populations of *Oscillatoria agardii*, *O. redekki* and *O. ornata* occurred, it was tempting to account for the almost total absence of desmids from the study lakes as a consequence of the excretion of inhibitory organic substances. Desmids were, however, found to grow quite satisfactorily in the filtrate from these blue-green populations, and as explained elsewhere, (p. 202), lack of water circulation was the most likely cause for their absence.

Lund (1971) questions whether inhibition by other algae plays a part in desmid ecology, having found from the analysis of 25 years of continuous sampling in the N. and S. basins of Windermere that in the latter basin, in which blue-greens are much more abundant, on the whole desmids are more abundant in this basin too. Thus in this, the best studied of all lakes, he discounts any suggestion that declines in desmid abundance can have ever been related to the abundance of other algal groups.

Relevant to the topic of the relationship between desmids and dissolved organics is Palmer's (1969) composite rating of algae tolerating organic pollution. His list is headed by the phytoflagellate *Euglena viridis* and second is the diatom *Nitzschia palea*. Twenty fifth in his rating is *Closterium acerosum* and 70th *C. leibleinii*. No indication is given as to why these forms have such tolerances. Confirmation of the tolerance of *Closterium* species to eutrophication is to be found in the study of seasonal succession and cultural eutrophication in a north temperate lake by Casterlin and Reynolds (1977) who found this genus to be a perennial component of its plankton showing peaks of abundance in July–August and September–October.

The high tolerance of *Closterium* species would seem to be confirmed by Palamar-Mordvintseva and Khisoriev (1979) who report that *C. acerosum* fac. *acerosum*, *C. peracerosum* and *C. moniliferum* reached significant numbers in the presumably highly organic waters in a purifying system at Dushanbe and in the Kafirnignan river. Also important were *C. botrytis* var. *botrytis* and *C. margaritaceum*. All of these desmids are said to play an important role in the processes of purification.

There would appear to be no evidence relevant to the other side of the coin—whether desmids excrete substances inhibitory to the growth of other algae or microorganisms. However, Soeding, Dorffling and Mix (1976) have shown, with the aid of an improved and more sensitive method for testing for bactericidal substances by paper chromatography that two bactericidal substances were found in ether extracts from bacteria-free cultures of *Cosmarium impressulum*, and two additional ones in extracts from the culture medium. These were active in their turn either against some, or all, of the bacteria tested. However, if the culture in which *C. impressulum* had been growing was innoculated with a test bacterium, no bactericidal effect was ever clearly demonstrated. Moreover, the positive, but very small, quantities of bactericidal substances produced (not on every occasion) by the desmid could not be conclusively implicated as the cause for the fact that epiphytic bacteria do not normally grow on *C. impressulum*.

DESMID GROWTH RATES AND POPULATION MAXIMA

Moss (1973b) in his experimental studies, has found that those desmids which are presumed to be HCO_3 (or low concentration of free CO_2 utilizers) have generally higher growth rates (0.50 doublings/day) than the non-users (0.30 doublings/day) (see his Fig. 3b). It is of considerable interest to note that in the Lund Tube experiments in which a large population of the eutrophic *Staurastrum pingue* developed (see p. 220), Reynolds writes that 'the maximum represented the culmination of a fairly consistent rate of increase over the preceeding 5 weeks: this works out to a mean doubling time of one every 2.52 days (about 0.40 doublings/day).

As pointed out by Lund (1964), the lower growth rates of oligotrophic species, as compared with eutrophic algae, may be of survival value. Thus, he argues, the faster a species grows the greater, it must be presumed, will be its nutrient demand. Thus, if a species with a high intrinsic growth rate occurs in a lake where nutrients are depleted, the cells may beeome nutrient starved and so die. Clearly, a eutrophic species in an oligotrophic lake would be at a competitive disadvantage, other things being equal, with the resident oligotrophic species. As Moss (1972) states, although oligotrophic species do occur quite commonly in eutrophic lakes they rarely, if ever, out-compete them. If, however, nutrient conditions change suddenly, as for example as a consequence of eutrophication, natural or artificial (see p. 220 and p. 222), then even the eutrophic species of desmids may rapidly attain ascendency.

However, the eutrophic desmids growing in eutrophic lakes do not normally become dominant because of their presumed much lower intrinsic growth rates than other phytoplankters in the community in which they occur. For example, Reynolds (1973) found 5 desmid species in the plankton of the eutrophic Crose Mere, Shropshire, which is rich in all major ions and nutrients (Gorham 1957), but they always formed a subsidiary element of its plankton. For much of the year, the lake supports large populations of diatoms, and then blue-green algae, appropriate to their seasons. Indeed in eutrophic lakes the contribution of desmids to the total phytoplankton biomass is, except under unusual circumstances (p. 221), very small but in oligotrophic lakes (Croome and Tyler 1973) it can be quite a significant component of the total biomass. In most lakes (and rivers) where they occur desmids are present in numbers of less than 10 cells/ml compared with often 1000's or 10 000's cells/ml of microflagellates, or even blue-greens, and occasionally diatoms.

Occasionally, however, desmids do 'take off' and as noted already (p. 221) under exceptional conditions large populations dominating the phytoplankton community result. One such case has been recorded from Green Lake, Isanti Co., Minnesota (Brook 1971) where 400 cells/ml of *Staurastrum pingue* and *S. chaetoceras* were counted. This was considered to be

Q

an 'after bloom' following extensive treatment of the lake with $CuSO_4$ in an attempt to control extensive blooms of blue-green algae. The largest population of desmids ever recorded would seem to be that from Siblybae Reservoir, Nr Liscard, Cornwall (Lack, personal communication). In this a bloom of alarming proportions was reported to Dr T. Lack of the Water Research Centre by the S.W. Water Authority, and from a sample dated 13 May 1973 he counted in it 30 082 cells/ml of *Arthrodesmus incus* var. *ralfsii* (Fig. 78: 1, 2). Lack has provided the data in Table 11 relevant to this bloom.

Table 11. Data for desmid bloom at Siblybae Reservoir.

Date	Depth	Desmid cells/ml	Chlorophyll 'a' mg/l
13 May	0 m	30 082	324
	3 m	26 148	330
15 May	0 m	36 600	—
	3 m	—	—
22 May	0 m	27 540	284
	3 m	27 500	356

Records of dissolved oxygen during the day during this period showed 110% saturation (temperature was about 10°C) between 0–15 m; pH varied between 8.1 and 8.3 (cf. p. 219). The only suggested cause for this massive blooming of a desmid was that arable land surrounding the reservoir had earlier been enriched by chicken manure.

Another record of a lake with a large desmid population is given by Swale (1968) in her 1963–66 study of the phytoplankton of Oak Mere, Cheshire, England. She notes that *Closterium gracile* increased rapidly from June 1964 to become a conspicuous element of the plankton throughout the subsequent autumn and winter. A maximum of 9400 cells/ml were recorded at the end of August but around 4000 cells/ml were present continuously until mid-December. *C. acutum* var. *variabile* became conspicuous at about the same time as *C. gracile* and attained a maximum of 13 800 cells/ml in October but it failed to persist as did the latter into the following spring, and indeed virtually disappeared in November. Swale records considerable fungal parasitism by chytrids of the *C. acutum* population at the end of August 1964 (cf. p. 236).

In addition to the large population densities already quoted, very large populations have been recorded from two eutrophic Tasmanian lakes, Sonell and Coescent (Tyler, personal communication) in which the following maxima have been found:

Euastrum cuspidatum var. *goyazense* 350 cells/ml

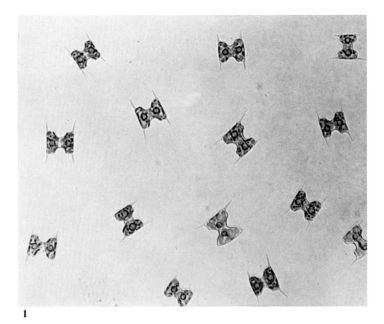

1

2

Fig. 78. A very small part of the largest population of desmids recorded; from Siblybae Reservoir, Cornwall, England (30 000 cells/ml). The lower photomicrograph (2) shows the mucilage investment surrounding each cell by mounting in Indian ink. (Photomicrographs by Dr Hilda Canter-Lund.)

Spondylosium planum 60 cells/ml
Cosmarium contractum var. *minutum* 18 cells/ml
Staurastrum pingue 35 cells/ml
Staurastrum excavatum 800 cells/ml

From a small alkaline pond in Spain De Emiliani (1973) records early October populations of 158 cells/ml at the surface, and 594 at 2 m of *Closterium acutum*, and for the same depths 39 and 148 cells/ml, respectively, of *Cosmarium tenue* var. *minus*.

THE PERIODICITY OF DESMIDS

Largely it is believed, because of their relatively small numbers, there is much less known about the periodicity of desmids than of other algal groups, such as diatoms or blue-greens. By far the most detailed accounts of desmid periodicity are the studies of Canter and Lund (1966) and Lund (1971). The latter paper provides details of the periodicity of *Cosmarium abbreviatum*, *C. contractum* and *Staurastrum lunatum* in the N. and S. Basins of Windermere in the English Lake District over a period of 25 years. The study shows that these desmids in general are most abundant in the second half of the summer, their numbers beginning to increase in May with maxima between June and August, or September and October, depending on the species.

Thus for *C. contractum* the maximum is reached in September or October, while for *C. abbreviatum* it seems to be more variable, though commonly it is between June and August. It can, however, be as early as May and there can be significant increases of this species in October or even November. *S. lunatum* has its maximum usually in July or August but it has been in June and also in September. Lund remarks that overall, the periodicity of *S. lunatum* is similar to that of *C. contractum* though the time of maximum abundance of the latter is most frequently about a month later than of the former.

With reference to their depth distribution in Windermere all three species showed increases in the epilimnion which more or less coincides in depth with the euphotic zone, though when the lake is isothermal distribution then becomes random.

Growth experiments of the three desmids under lake conditions were carried out in Chu 10 culture medium (Chu 1942) with added soil extract by suspending the desmids in glass bottles in the lake at 0.5, 1.0, 2.0 and 4.0 m depth. At 6.0 m growth was usually very slight. In agreement with the observations of the dynamics of the populations of the three species in the lake, at 0.5, 1.0 and 2.0 m the desmids began to show significant increases in growth from April or May, but usually began to decrease in August. Also the growth rate of *C. contractum* was, except for one occasion, less than that of *C. abbreviatum* or *S. lunatum*. Its slower growth rate helps in part to explain why this desmid reaches its maximum later than that of the other two species; though as Lund points out, since potentially these two species are able to grow faster at all times and over the whole photic zone, 'there seems to be no reason why they should not be as abundant than they actually are in autumn'. Again the difference would seem to hinge on the competitive advantage of a slow rate of growth (cf. p. 231). As Lund explains, since a late maximum will occur at a time when stratification is breaking down, or has been lost, the population will be dispersed through a greater depth than if the maximum was reached earlier. He quotes as an example the situation in 1953 when isothermal conditions extended down to 20 m and *C. contractum* was it its maximum.

In the S. Basin of Windermere the volume of the lake from surface to 20 m is about 75% of the total. During the maximum of *S. lunatum* and *C. abbreviatum*, usually July–August, the epilimnion extends only to 5–10 m, therefore these desmid populations occur only in some 25–45% of the S. Basin and 15–40% of the N. Basin. Consequently as mixing occurs and the metalimnion deepens, so the population per unit volume decreases, unless an increasing growth rate (as happens relatively for *C. contractum*) overcomes this dilution.

No attempt has been made to relate changes in the abundance of these 3 planktonic desmids to the changing seasonal conditions in Windermere's water chemistry. Lund does, however, point out that the curve describing their seasonal wax and wane parallels temperature rather than incident radiation.

In a study of a series of sub-tropical mountain lakes of low ionic content, in the Pokhara Valley, Nepal (Hickel 1973), in which desmids were sub-dominant to diatoms in the plankton, desmid maxima (2×10^5 cell/ml) were found in early December, in March and again in May. With reference to their depth distribution, in December desmids were most abundant at 2.5 m, while in March and May the maxima were at about 7 m.

The early studies of Ruttner (1930) in the Lunzer Untersee, a relatively cold alpine lake, showed that 3 *Staurastrum* species, 2 of *Staurodesmus* and *Closterium aciculare* increase during the summer months, though attaining their maximal abundance at different times in different years (1909–1912). As Hutchinson (1967) states, with reference to the *Staurastrum* species, it would seem reasonable to assume that each of those involved has slightly different ecological requirements and indicates that this can be demonstrated for the two commonest, *S. cingulum* (*S. paradoxum* in Ruttner 1930) and *S. luetkemuelleri* (cf. *S. pingue*).

Thus *S. cingulum* begins to increase at a slightly lower temperature than *S. luetkemuelleri*. In the hot summer of 1911 the latter was, however, much more abundant than *S. cingulum* at the beginning of September when the growth rate of *S. luetkemuelleri* was almost maximal, while that of *S. cingulum* was significantly lower than it had been earlier. During the 1911 summer the surface waters of the lake were continuously above 15°C from early July to mid-September. In 1912 temperatures above 15°C were never recorded and were much lower throughout September. However, *S. cingulum* increased in August more or less to the level of abundance of the preceeding year, while *S. luetkemuelleri* remained scarce.

In discussing the competitive relationships between the two species, Hutchinson (1967) states that it 'might be reasonable to suppose that if the temperature were to remain permanently above 15°C, *S. lütkemülleri* would displace *S. cingulum*, whereas if the temperature remained permanently between 12 and 13°C *S. cingulum* would displace *S. lütkemülleri*. In nature, however, before competition has proceeded very far, the temperature falls in the lake and doubtless other conditions later; both species decline and remain of little importance throughout the winter'.

In Crose Mere (Reynolds 1973 and p. 231) found the largest numbers of *Staurastrum cingulum* (4 cells/ml) in May 1967 and July 1968. *Closterium* species (*C. aciculare* var. *subpronum*, *C. acutum* and *C. tortum*) became most abundant in autumn following a slow rate of increase and frequently the populations (4–20 cells/ml) persisted through to following spring. In 1971, Reynolds reported a maximum of *C. aciculare* var. *subpronum* at the end of December. (cf. Youngman, Johnson and Farley 1976 and p. 237).

An almost completely unexplored field of desmid ecology is that of the epipelic community. There is however some data for changes in standing crop by Gruendling (1971) for Marion Lake, British Columbia, Canada. Again *Closterium* was the dominant desmid genus and both *C. gracile* and *C. intermedium* were at a minimum in January at densities of 1.05 cells $\times$ 10^3/cm^{-2}. The former achieved a maximum in October of 0.65 and the latter in August–September with 0.60 cells $\times$ 10^3/cm^{-2}.

FUNGAL PARASITISM OF DESMIDS

Lund (1971) suggests that the late maximum of *C. contractum* confers another advantage in that there is less likelihood of it being infected by fungal parasites which can cause severe depredations to desmid populations (Canter and Lund 1969). Thus in the 4-year period during which desmid parasites in Windermere were studied in detail *S. lunatum* (Fig. 79: 3, 4) suffered two periods when over 30% of the population were infected, whereas in the same period *C. contractum* (Fig. 79: 1, 2) had only one. The lower incidence of infection may be a temperature effect

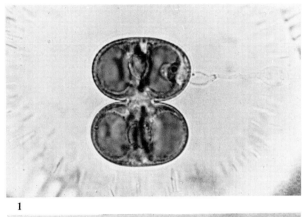

1

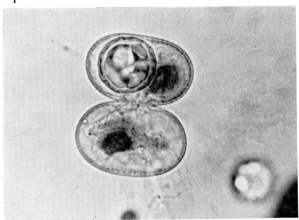

2

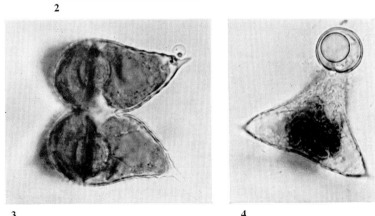

3   4

Fig. 79. Fungal parasitism of planktonic desmids in Windermere. 1 and 2 Stages in the infection of *Cosmarium contractum* Kirchn. 3 and 4 Stages in the infection of *Staurastrum lunatum* Ralfs. Note in 2 and 4 the cell contents have contracted dramatically due to fungal parasitism. (Photomicrographs by Dr Hilda Canter-Lund.)

since fungi grew less well in cold water than in warm; but it is also suggested that since many desmid parasites are unselective as to their desmid hosts there are less potentially infective desmids about when *C. contractum* reaches its maximum. Although fungal epidemics may be severe (Canter and Lund 1969) and affect healthy rapidly growing cells, they conclude that their effect on desmid populations as a whole does not seem to alter their normal seasonal periodicity. However, they do concede that many species would be more abundant almost every year in Windermere and other lakes were it not for such parasitism.

Studies at the Thames Water Authority's Farmoor Reservoir, 6 km west of Oxford (Youngman *et al.* 1976), during the development of a bloom of phytoplankton from November 1973 to March 1974 are of interest in that fungal parasitism was almost certainly implicated in the demise of the desmid component of the phytoplankton. During these observations the development and decline of 4 phytoplankters was observed. Of these the diatom *Asterionella formosa* was interrupted in December when 43.5% of its cells were parasitised by *Zygorhizidium affluens*, though as with succeeding populations of *Stephanodiscus astrea* and *S. hantzschii* growths of all 3 populations were finally limited by silica depletion. The fourth dominant phytoplankter was *Closterium peracerosum* which became a significant component in November, increasing slowly to reach a maximum of 14.9 cells/ml on 18 February 1974. In the first week of this month, however, it was noticed that 2% of cells were infected by a so far unidentified epiliotic, operculate chytrid. By 4 March 66% were infected and by this date the population was declining rapidly so that by 25 March only 0.3 cells/ml remained (see also Reynolds 1973, and Fig. 80; also Swale 1968 and p. 231.)

An August maximum of *Staurastrum pingue* in the Titisee was shown by Schulle (1970) to have declined from 16 cells/ml to almost zero following fungal parasitism by a chytrid fungus.

Reynolds (1940) found that the proportionate abundance of triradiate and biradiate facies of his 'Staurastrum paradoxum' (see p. 33) changed in relation to fungal infestation. Although Brook (1959a) is of the opinion that Reynolds was dealing with only one species, *S. chaetoceras*, Canter and Lund (1966) believe that two distinct taxa were involved. If this is a correct interpretation then the interesting implication is that interspecific competition between desmids is affected by fungal parasitism.

Although Canter (1949 a, b, c) has shown that many desmids occurring in the metaphyton are parasitised to quite a considerable extent from time to time, the impact of such infections on the dynamics of these populations is unknown but may well be a very significant factor at times. In her study of snow and ice algae in Alaska, Kol (1942) noted that the chytrid *Rhizophidium sphaerocarpum* parasitised the cells of the most widely occurring,

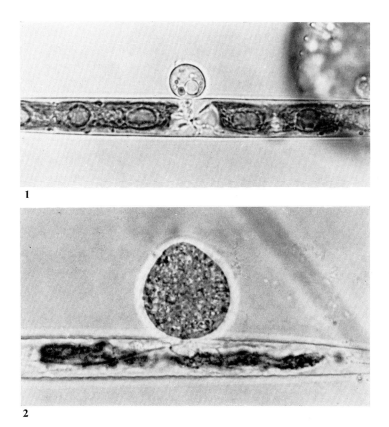

Fig. 80. Fungal parasitism of planktonic desmids. Stages in the infection and parasitism of *Closterium aciculare* Tuffen West in Crose Mere: 1 An early stage in the parasitism of a cell; 2 later stage in which cell contents have contracted and rhizoids from the fungus can be seen within the *Closterium* cell. (Photomicrographs by Dr Hilda Canter-Lund.)

and often very abundant cryobiont, *Ancylonema nordenskioldii* (see p. 209).

An extensive host-range investigation was undertaken by Cook (1963) to evaluate the host–parasite interactions as an indication of the possible taxonomic relationships of the hosts. Using the appressoria, or presence of germinating zoospores, as susceptibility to infection, Cook found that this was closely correlated with the morphological characteristics of the desmids. For example, *Closterium moniliferum* and related taxa, their susceptibility to several specific parasites supported the delimitation of these desmids on the basis of their vegetative and sexual morphology. In the case of the *C. acerosum–praelongum* complex resistance or susceptibility was more closely associated with sexual morphology than the vegetatitive morphological characteristics used as taxonomic criteria to delimit species. Moreover, the ability of some fungal parasites to infect only certain clones of *C. praelongum* var. *bevius* was correlated with morphological differences between the clones. Host–parasite reactions between 13 taxa of the genus *Micrasterias* and the fungus termed *Myzocytium II*, also provided a basis for the grouping of the species.

GRAZING ON DESMIDS BY THE AQUATIC FAUNA

The extent to which desmid abundance and their periodicity in the plankton is affected by the grazing of Crustacea, Rotifera or Protozoa is a virtually unknown and unexplored field. Lund (1971, p. 23) has remarked that in Windermere only the rotifer *Asplanchna priodonta* might affect seasonal cycles of planktonic desmids. Although large ciliated Protozoa might also consume them, Canter and Lund (1968) in examining the significance of Protozoa in controlling the abundance of phytoplankton make no reference to them in relation to desmids.

In an experimental lake study 0.5 m² polyethylene enclosures were filled with lake water, filtered through a 125 μm mesh net to remove the major zooplankton grazers; comparing the phytoplankton in these with that in similar bags containing lake water, with its natural complement of zooplankton, Porter (1972, 1973) found no detectable effect of the latter on the abundance of desmids after 4 days of experimental incubation. The significant effect was the suppression of small algae, primarily flagellates, and other nannoplankters and large diatoms. Examination of the guts of the herbivorous grazers *Daphnia galatea mendota* and *Diaptomus minutus*, revealed that desmids were rarely, or never, ingested. However, Fryer (1957) in a study of the food of some cyclopoid copepods states that *Acanthocyclops bisetosus* will ingest *Cosmarium* cells, but when the cell walls of the desmid are not ruptured the cells pass out apparently unharmed. In a study of the food of copepods in the man-made L. Kayanji, Nigeria, Clarke (1978) found as many as 6 unidentified *Staurastrum* cells in the gut of one *Thermocyclops neglectus*.

In small-scale culture experiments it can be shown that several different zooplankters can ingest, and digest, various desmid species. Thus Lack (pers. comm.) has demonstrated that *Daphnia hyalina* will ingest considerable numbers of desmids including *Staurastrum chaetoceras*. A series of experiments with *D. magna* (pers. obs.) have shown that *Cosmarium contractum*, *C. reniforme*, and *Closterium peracerosum* can all be ingested and in turn digested by this cladoceran, to the extent that all that remains in the cultures after 3 or 4 days are empty cells. Although observations point to the conclusion that desmids are apparently unaffected by zooplankton grazers, their cells not appearing in their guts, this may be not because they cannot be eaten and digested, but rather because they are present in small numbers in comparison with other phytoplankters, and so their chances of being consumed are very low: i.e. they normally occur in densities of much less than 10 cells/ml as compared with many 1000 cells/ml in the ease of many other planktonic algae.

In studies on the feeding of ciliates on freshwater benthic algae, in the slow sand filters of waterworks (Brook 1954), very occasionally desmids were seen to have been ingested, although this flora was dominated by diatoms which the ciliates consumed in large numbers.

During observations on the feeding of aquatic insect larvae on algae of the metaphyton desmids have been found from time to time in the guts of mayfly nymphs (*Leptophlebia vespertina*, Brook 1955). In these studies it was also found that diatoms and filamentous green algae were their most important source of food, and the few desmids that were observed seem to pass through the guts of these invertebrates unharmed.

THE ROLE OF DESMIDS IN ECOSYSTEMS AND CONCLUSIONS

In the introduction to this chapter it was stated that one of the major aims confronting the desmid ecologist was to attempt to discover the role of these algae in the functioning of the ecosystems in which they occur. However, in the absence of any clear indication that they contribute to the food supply of any component of the aquatic fauna, though obviously 'appreciated' by fungal parasites, their role is obscure. Indeed all that can be said is that they are one component amongst a great diversity of primary producers which capture the sun's energy in aquatic ecosystems and by the same token must be involved in nutrient cycling.

The concluding words about them must be those of Dr John Lund (to whom this book is dedicated in appreciation of a lifetime of work devoted to the ecology of freshwater algae); '. . . desmids are very attractive green algae . . . but because they never upset fishing or offend the public, interest in them is considered to be "academic" and without much "cost benefit". However, they are delicate indicators of environmental change in ways we do not clearly understand and there is need for more ecological research on these allegedly unimportant plants. This the freshwater biologist may know, but it can be another matter to convince the administrator' (Lund 1973).

BIBLIOGRAPHY

ACTON E. (1916) Studies on nuclear division in desmids I. *Hyalotheca dissiliens* (Sm.) Bréb. *Ann. Bot.* **30**, 379–382.

ALCORN G.D. (1975) A new staining technique for Desmids. *Stain Technol.* **10**, 107–108.

ALLEN M.A. (1958) The biology of a species-complex in *Spirogyra*. PhD Thesis, Indiana University, Bloomington, 240 pp.

ANDRESON N.A. (1973) A note on the nomenclature of desmids. *Beih. Nova Hedwigia*, **42**, 27–31.

BACKHAUS D. & KEMBALL A. (1978) Gewassergüte verhältnisse and Phytoplankton entwicklung in Hochrhein, Oberrhein und Necker. *Arch. Hydrobiol.* **82**, 166–206.

BAILEY-WATTS A.E. (1974) The algal plankton of Loch Leven, Kinross. *Proc. roy. Soc. Edinb.* **74**, 135–156.

BAKER A.L. (1973) Microstratification of phytoplankton in three Minnesota lakes. PhD Thesis, University of Minnesota.

BAKER A.L. & BROOK A.J. (1971) Optical density profiles as an aid to the study of microstratified phytoplankton populations in lakes. *Arch. Hydrobiol.* **69**, 214–233.

BARG T. (1942) Beitrag zur Cytomorphologie der Desmidiaceen. *Arch. f. Protistenk.* **95**, 391–432.

BARTELS P.G. & HOWSHAW R.W. (1968) Cylindrical structures in the chloroplasts of *Sirogonium melanosporum*. *Planta*, **82**, 293–298.

BEAMS H.W. & KESSEL R.G. (1986) The Golgi Apparatus: structure and function. *Int. Rev. Cytol.* **23**, 209–276.

BEHRE K. (1956) Die Algen besiedlung einiger Seen um Bremen und Bremerhaven. *Ver. Inst. Meersk. Bremerhausen*, **4**, 221–283.

BEHRE K. (1966) Zur Algensoziologie des Süsswassers. *Arch. Hydrobiol.* **62**, 125–164.

BEIJERINCK, W. (1926) Over veuspreiding en periodiciteit van de zoctwaterwieren in Deutsche heideplassen. *Verr. Ken. Ned. Acad. Weteusch. Afd Natuurk. Tweede Sect.* **25**, 5–211.

BENDIX S. (1960) Phototaxis. *Bot. Rev.* **26**, 145–208.

BENNETT A.W. (1889) A hybrid desmid. *Ann. bot.* **4**, 171.

BENNETT A.W. (1892) Spore-like bodies in *Closterium*. *Ann. Bot.* **6**, 150–152.

BERGGREN S. (1871) Alger fran Grönlands inlandis. Kongl. *Vetensk.-Akad. Förhandl.*, **28**, 293–296.

BERLINER M.D. & WENC K.A. (1976a) Osmotic pressure effects and protoplast formation in *Cosmarium turpinii*. *Microbios*. **2**, 39–45.

BERLINER M.D. & WENC K.A. (1976b) Protoplasts of *Cosmarium* as a potential protein source. *Appl. Envir. Micribiol.* **32**, 436–437.

BERNSTEIN E. (1960) Synchronous division in *Chlamydomonas moewusii*. *Science*, **131** 1528–1529.

BOCHER T. (1949) Climate, soil and lakes in continental West Greenland in relation to to plant life. *Meddr. Grønland*, **147**, 1–63.

BHARATI, S.G. (1965) *Micrasterias euastriellopsis* sp. nov. an interesting new species of the Genus *Micrasterias*. *Rev. Algol, N.S.* **8**, 43–45.

BICUDO C.E.M. (1973) Typification of the generic desmid name *Arthodesmus*. *Beih. Nova Hedwigia*, **42**, 33–38.

BICUDO C.E.M. (1975) Polymorphism in the desmid *Arthodesmus mucronulatus* and its taxonomic implications. *Phycologia*, **14**, 145–148.

BICUDO C.E.M. (1976) *Prestcottiella*, a new genus of asymmetrical desmid. *J. Phycol.* **12**, 22–24.

BICUDO C.E.M. & CARVALHO L.M. (1969) Polymorphism in the desmid *Xanthidium regulare* and its taxonomic implications. *J. Phycol.* **5**, 369–375.

BICUDO C.E.M. & SORMUS L. (1972) Polymorphism in the desmid *Micrasterias laticeps* and its taxonomic implications. *J. Phycol.* **8**, 237–242.

BICUDO C.E.M. & SORMUS L. (1977) Typification of the generic desmid name *Micrasterias* Desmidiaceae. *Taxon*, **26**, 322–323.

BIEBEL P. (1964) The sexual cycle of *Netrium digitus*. *Am. J. of Botany*, **51**, 697–704.

BIEBEL P. (1975) Morphology and life cycles of saccoderm desmids in culture. *Beih. Nova Hedwigia*, **42**, 39–57.

BIEBEL P. & REID R. (1965) Inheritance of mating types and zygospore morphology in *Netrium digitus* var. *lamellosum*. *Proc. Pa. Acad. Sci.* 39, 134–137.

BIEBL R. & ROSSI-PILLHOFER W. (1954) Die Anderung der chemischen Resistenz pflanzlichen Plasmen mit dem Entwickwingszustand. *Protoplasma*, **44**, 113–135.

BLACKBURN S.I. & TYLER P.A. (1980) Conjugation, Germination and Meiosis in *Micrasterias mahabuleshwarensis* Hobson (Desmideaceae). *Brit. Phycol. J.* **15**, 83–93

BLAND R.D. & BROOK A.J. (1974) The spatial distribution of desmids in lakes in northern Minnesota, U.S.A. *Freshwat. Biol.* **4**, 543–556.

BORGE O. (1892) Subfossila sotvattensalger från Gotland. *Bot. Notiser* (1892), 55–58.

BORGE O. (1895) Über subfossile Süsswasseralger från Gotland. *Bot. Centralbl.*, **63**, 45–63.

BORGE O. (1896) Nachtrag zu der subfossilen Desmidiaceen flora Gotlands. *Bot. Notiser* (1896), 111–113.

BORGE O. (1896) Uber die Variabilität der Desmidiaceen. *Ofr Kongl. Ventensk Akad. Forh* 289–294.

BORGE O. (1899) Süsswasseralgen von Franz Josefs-Land, gesammelt von der Jakson-Harmsworth'schen Expedition. *Kongl. Vetensk.-Akad. Forhandl.*, **56**, 751–766.

BOURRELLY P. (1964) Une nouvelle coupure générique dans la famille des Desmidiées, le genre *Teilingia*. *Rev. Algol. N.S.* 7, 187–191.

BOURRELLY P. (1966) Les Algues d'eau douce, Tome I—algues vertes Paris.

BOURRELLY P. (1977) *Gonatozygon chadefaudii*, nouvelle espece de la Guyana, (Zygophyceae: Closteriaceae). *Rev. Algol. N.S.* **12**, 3–8.

BRANDHAM P.E. (1964) Cytology, sexuality and mating type in culture of certain desmids, Vols. I and II. PhD Thesis, U. of London.

BRANDHAM P.E. (1965a) Formation, division and conjugation of multinucleate cells of desmids. *Phycologia*, **5**, 45–52.

BRANDHAM P.E. (1965b) Some new chromosome counts in the desmids. *Brit. Phycol. Bull.* **2**, 451–455.

BRANDHAM P.E. (1965c) The effect of temperature on the radial symmetry of *Staurastrum polymorphum*. *J. Phycol.* **1**, 55–57.

BRANDHAM P.E. (1965d) The occurrence of parthenospores and other haploid resistant spores in desmids. *Trans. Amer. Microsc. Soc.* 74, 478–484.

BRANDHAM P.E. (1965e) Polyploidy in Desmids. *Canad. J. Bot.* **43**, 405–417.

BRANDHAM P.E. (1967a) Three new desmid taxa from West Africa, including two asymmetrical forms. *Brit. Phycol. Bull.* **3**, 189–193.

BRANDHAM P.E. (1967b) Time-lapse studies of conjugation in *Cosmarium botrytis* I. Gamete fusion and spine formation. *Rev. Algol. N.S.* **8**, 312–316.

BRANDHAM P.E. (1967c) Time-lapse studies of conjugation in *Cosmarium botrytis* II. Pseudoconjugation and an anisogamous mating behaviour involving chemotaxis. *Canad. J. Bot.* **45**, 483–493.

BRANDHAM P.E. & GODWARD M.B.E. (1963) Mating types and meiosis in Desmids. *Brit. Phycol. Bull.* **2**, 280–281.

BRANDHAM P.E. & GODWARD M.B.E. (1964) The production and inheritance of the haploid form in *Cosmarium botrytis*. *Phycologia*, **4**, 75–83.

BRANDHAM P.E. & GODWARD M.B.E. (1965a) The inheritance of mating type in desmids. *New Phytol*, **64**, 428–435.

BRANDHAM P.E. & GODWARD M.B.E. (1965b) Mitotic peaks and mitotic time in the Desmidiaceae. *Arch. Mikrobiol.* **51**, 393–398.

BRANDHAM P.E. & GODWARD M.B.E. (1965c) Ultrastructure of the nucleus of the green algae *Closterium ehrenbergii*. *J. Roy. microsc. Soc.* **84**, 499–507.

BRIGHT R.C. (1968) Surface-water chemistry of some Minnesota lakes, with preliminary notes on diatoms. *L.R.C. U. of Minn., Interim Report No.* **3**, 59 pp.

BROOK A.J. (1954) A systematic account of the phytoplankton of the Blue and White Nile. *Ann. Mag. Nat. Hist. Ser.* 12, **7**, 648–656.

BROOK A.J. (1955) The aquatic fauna as an ecological factor in studies of the occurrence of freshwater algae. *Rev. Algol. N.S.* **1**, 141–145.

BROOK A.J. (1958a) Desmids from the plankton of some Irish loughs. *Proc. Roy. Irish. Acad.* **59B**, 71–91.

BROOK A.J. (1958b) Changes in the phytoplankton of some Scottish hill lochs resulting from their artificial enrichment. *Verh. int. Ver. Limnol.* **13**, 298–305.

BROOK A.J. (1959a) The status of Desmids in the plankton and the determination of phytoplankton quotients. *J. Ecol.* **47**, 429–445.

BROOK A.J. (1959b) *Staurastrum paradoxum*, Meyen and *S. gracile*, Ralfs in the British freshwater plankton and a revision of the *S. anatinum*-group of radiate desmids. *Trans. Roy. Soc. Edinb.* **58(3)**, 589–628.

BROOK A.J. (1959c) The published figures of the desmid *Staurastrum paradoxum*. *Rev. Algol. N.S.* **4**, 239–255.

BROOK A.J. (1965) Planktonic algae as indicators of lake types with special reference to the Desmidiaceae. *Linnol. Oceanogr.* **10**, 403–411.

BROOK A.J. (1971) The phytoplankton of Minnesota Lakes—a preliminary survey. *Wat. Res. R. Center., U. of Minn. Bull.* **36**, 12 pp.

BROOK A.J. & HINE A.E. (1966) A population of *Staurastrum freemanii* from the Central Highlands of New Guinea. *J. Phycol.* **2**, 66–73.

BROOK A.J. & HOLDEN A.V. (1957) Fertilization experiments on Scotland freshwater lochs I. Loch Kinardochy. *Sci. Invest. Freshwat. Fish. Scot.* **No. 17**, 30 pp.

BROOK A.J. & RZÖSKA J. (1954) The influence of the Gebel Aulyia Dam on the development of Nile plankton. *J. Animal Ecol.* **23**, 101–114.

BROOK A.J. & WOODWARD W.B. (1956) Some observations on the effects of water inflow and outflow on the plankton of small lakes. *J. Animal Ecol.* **25**, 22–35.

BROOK A.J., FOTHERINGHAM J., BRADLY J. & JENKINS A., (1980) Barium accumulation by desmids of the genus *Closterium* (Zygnemaphyceae). *Brit. Phycol. J.* **15** (in press).

BROWN R.M. Jr, LARSON D.A. & BOLD H.C. (1964) Airborne algae; their abundance and heterogeneity. *Science N.Y.* **143**, 583–585.

CANTER H.M. (1949a) Studies on British Chytrids VI. Aquatic Synchitriaceae. *Trans. Brit. Mycol. Soc.*, **32**, 69–94.

CANTER H.M. (1949b) On *Apharomycopsis bacillariacearum* Scherffel, *A. desmidiella* n.sp., and *Ancylistes* spp. in Great Britain. *Trans. Brit. Mycol. Soc.*, **32**, 162–170.

CANTER H.M. (1949c) Studies on British Chytrids VII. On *Phytochytrium mucronatum*. *Trans. Brit. Mycol. Soc.*, **32**, 236–240.

CANTER H.M. (1954) Fungal parasites of the phytoplankton III. *Trans. Brit. Mycol. Soc.*, **37**, 111–132.

CANTER H.M. & LUND J.W.G. (1966) The periodicity of planktonic desmids in Windermere, England. *Verh. int. Ver. Limnol.* **16**, 163–172.

CANTER H.M. & LUND J.W.G. (1968) The importance of Protozoa in controlling the abundance of planktonic algae in lakes. *Proc. Linn. Soc. Lond.* **179**, 203–219.

CANTER H.M. & LUND J.W.G. (1969) The parasitism of planktonic desmids by fungi. *Öst. bot. Zeit.* **116,** 351–377.

CARTER N. (1919a) Studies of the chloroplasts of desmids I. *Ann. Bot.* **34,** 265–255.

CARTER N. (1919b) Studies of the chloroplasts of desmids II. *Ann. Bot.,* **33,** 295–304.

CARTER N. (1920a) Studies on the chloroplasts of desmids III. *Ann. Bot.,* **34,** 265–285.

CARTER N. (1920b) Studies on the chloroplasts of desmids IV. *Ann. Bot.,* **34,** 303–319.

CASTERLIN M.E. & REYNOLDS W.W. (1977) Seasonal succession and cultural eutrophication in a north temperate lake. *Hydrobiologia* **54,** 99–108.

CEDERGREN G. (1932) Die Algenflora der Provinz Harjedalen, *Ark. Bot.* **25A,** 1–107.

CEDERGREN G.R. (1938) *Cosmarium*–und chlorophyceereiche Algen–assoziation der Lacustrinen Serie. *Bot. Notiser,* **91,** 911–12.

CHADEFAUD M. (1936) Le cytoplasme des algues vertes et des algues brunes. *Rev. Algo* . **8,** 1–286.

CHANDLER D.C. (1937) Fate of typical lake plankton in streams. *Ecol. Monogor.* **7,** 445–479.

CHARDARD R. (1964) Étude de l'ultrastructure de deux algues Zygophycées. *Mesotaenium cladariorium* et *Penium margaritaceum. Rev. Cytol. Biol. Végét.* **27,** 77–93.

CHARDARD R. (1965) Nouvelles observations sur l'infrastructure de deux Algues Desmidiales *Cosmarium lundellii* et *Closterium acerosum. Rev. Cytol. Biol. Végét.* **28,** 15–30.

CHARDARD R. (1966) Aspects infrastructure aux de l'autvantagonisme chez deux Desmidiales. *Rev. Cyt. Biol. Végét.* **29,** 336–356.

CHARDARD R. (1970) Action des solutions hypertonique sur l'ultrastructure d'un algue verte, *Cosmarium lundelli. Bull. soc. bot. France. Mém Colloque de cytologic expt.* **10,** 59–81.

CHARDARD R. (1972) Production de protoplastes d'algues par un procédé physique. *C.R. Acad. Sci. (Paris),* **274,** 1015–1018.

CHARDARD R. (1973) Action de la lumière et de l'obscurité sur la structure des plastes de *Cosmarium lundellii* et de *Closterium acerosum* (Desmidiaceae). *Beih. Nova Hedwigia,* **42,** 59–81.

CHARDARD R. (1974) La paroi de *Cosmarium lundellii* ultrastructure et essai de localization cytochemique des constituents. *C. R. Acad. Sci.* (Paris), **278,** 609–612.

CHARDARD R. (1977) La secretion de mucilage chez quelques desmidiales I. Les pores. *Protistologica,* **13,** 241–251.

CHARDARD R. & ROUILLER C. (1957) L'ultrastructure de trois algues Desmidicées. Etude au microscope électronique. *Rev. Cytol. Biol. Végét.* **18,** 153–178.

CHU S.P. (1942) The influence of the mineral composition of the medium on the growth of planktonic algae. I Methods and culture media. *J. Ecol.* **30,** 284–325.

CHU S.P. (1943) The influence of the mineral composition of the medium on the growth of planktonic algae II. The influence of the concentration of inorganic nitrogen and phosphate phosphorus. *J. Ecol.* **31,** 109–148.

CLARKE N.V. (1978) The food of adult copepods in L. Kainji, Nigeria. *Freshwat. Biol.* **8,** 321–326.

COESEL P.F.M. (1974) Notes on sexual reproduction in Desmids I. Zygospore formation in nature (with special reference to some unusual records of zygotes). *Acta. Bot. Neerl.* **23,** 361–368.

COESEL P.F.M. (1975) The relevance of desmids in the biological typology and evaluation of fresh waters. *Hydrobiol. Bull. (Amsterdam),* **9,** 93–101.

COESEL P.F.M. (1977) On the ecology of desmids and the suitability of these algae in monitoring the aquatic environment. *Hydrobiol. Bull. (Amsterdam),* **11,** 20–21.

COESEL P.F.M. & TEIXEIRA R.M.V. (1974) Notes on sexual reproduction in Desmids II. Experiences with conjugation experiments in uni-algal cultures. *Acta. Bot. Neerl.* **23,** 603–611.

COESEL P.F.M. & KWAKKESTEIN R. & VERSCHOOR A. (1978) Oligotrophication and eutrophication tendencies in some Dutch moorland pools, as reflected in their desmid flora. *Hydrobiologia,* **61,** 21–31.

COMPÈRE P. (1976a) The typification of the genus *Arthrodesmus*. *Taxon*, **25**, 359–360.

COMPÈRE P. (1976b) *Bourellyodesmus*, nouveau genre de Desmidiaceae. *Rev. Algol. N.S.*, **11**, 339–342.

COOK P.W. (1963a) Variation in vegetative and sexual morphology among the small curved species of *Closterium*. *Phycologia*, **3**, 1–18.

COOK P.W. (1963b) Host-range studies of certain phycomycetes, parasitic on desmids. *Amer. J. Bot.* **50**, 580–588.

COOK P.W. (1967) A cytochemical study of acid phosphatase and nuclease activity in the plastid of *Closterium acerosum*. *J. Phycol.* **3**, 24–30.

COUTÉ A. & RINO J.A. (1975) Structure de la membrane de *Cosmarium anthoporum* Couté et Rouss. *Protistologica*, **11**, 75–81.

CROASDALE H. (1955) Freshwater algae of Alaska I. Some desmids from the interior Part I. *Farlowia*, **4**, 513–565.

CROASDALE H. (1956) Freshwater algae from Alaska I. Some desmids from the interior, Part 2. *Alcinotaenium, Micrasterias* and *Cosmarium*. *Trans. Am. microsc. Soc.* **75**, 1–70.

CROASDALE H. (1957) Freshwater algae from Alaska I. Some desmids from the interior, Part 3. *Cosmariae* concluded. *Trans. Am. microsc. Soc.* **76**, 116–158.

CROASDALE H. (1962) Freshwater algae of Alaska III. Desmids from the Cape Thompson area. *Trans. Amer. Microsc. Soc.* **81**, 12–42.

CROASDALE H. (1965) Desmids from Devon Island N.W.T. Canada. *Trans. Amer. Microsc. Soc.* **84**, 301–355.

CROASDALE H. (1973) Freshwater algae of Ellesemere Island N.W.T. *Nat. Mus. Canada, Publ. Bot.* **3**, 1–131.

CROASDALE H. & SCOTT A.M. (1976) New or otherwise interesting desmids from N. Australia. *Nov. Hediwigia*, **27**, 501–595.

CROOME R.L. & TYLER P.A. (1973) Plankton populations of Lake Leake and Toems Lake–Oligotrophic Tasmanian lakes. *Br. phycol. J.* **8**, 239–247.

DE EMILIANI, M.O.G. (1973) Fitoplancton de la laguna del Vilá (Gerona, España). *Oecologia aquatica*, **1**, 107–155.

DOBBERSTEIN B. (1973) Einige Untersuchungen zur Sekundärwand bildung von *Micrasterias denticulata* Bréb. *Beih. Nova Hedwigia*, **42**, 83–90.

DOBBERSTEIN B. & KIERMAYER O. (1972) Das Auftreten eines besordern Type von Golgi —vesikeln wahrend der Sekundärwand bildung von *Micrasterias denticulata* Bréb. *Protoplasma*, **75**, 185–194.

DODGE J.D. (1963) The nucleus and nuclear division in the Dinophyceae. *Arch. Protistenk.* **106**, 442–452.

DORSCHEID T. & WARTENBERG A. (1966) Chlorophyll als photoreceptor bei der Schurachlichebewegung des *Mesotaenium*-chloraplasten. *Planta*, **70**, 187–192.

DRAWERT H. (1966) Der mikrospektralphotometrische nachweis von Chlorophyll b in Closterien and einige Mesotaeniceen *Ber. Deutsch. bot. Ges.* **19**, 403–407.

DRAWERT H. & KALDEN G. (1967) Licht und elektronmikroskopische Untersuchungen an Desmidiaceen XIII Mitteilung: Der Gestaltwandel des Nucleolus im Interphasekern von *Micrasterias rotata*. *Mitt. Staatsinst. Allg. Bot. Hamburg*, **12**, 21–27.

DRAWERT H. & METZNER-KÜSTER I. (1961) Licht und elektronmikroskopische Untersuchininigen au Desmidiaceen I. Zellwand—und Gallerstrukturen bei einigen Arten. *Planta*, **56**, 213–228.

DRAWERT H. & MIX M. (1961a) Licht und elektonmikroskopische Untersuchungen an Desmidiaceae II. Hüllgallerte und Schleimbildung bei *Micrasterias, Pleurotaenium* und *Hyalotheca*. *Planta*, **56**, 237–261.

DRAWERT H. & MIX M. (1961b) Licht und elecktronmikroskopische Untersuchungen an Desmidiaceen III. Der Nucleolus im Interphasekern von *Micrasterias rotata*. *Flora*, **150**, 185–190.

DRAWERT H. & MIX M. (1962) Licht und elektronmikroskopische Untersuchungen an Desmidiaceen X. Mitt Beitrage zur Kenntnis der Hautung von Desmidiaceen. *Arch. Mikrobiol.* **42**, 96–100.

DRAWERT H, & MIX M. (1963) Licht und elektronmikroskopische Untersuchungen an Desmidiaceen XI. Die Struktur von Nucleolus und Golgi—Apparat. bei *Micrasterias denticulata*. *Portug. Acta. Biol.* **7**, 17–28.

DROOP M.R. (1962) Organic micronutrients. In *Physiology and Biochemistry of the Algae* (Ed. R. Lewin). Academic Press, N.Y.

DUBOIS-TYLSKI T. Le cycle de *Closterium moniliferum* 'in vitro'. *Mem. Soc. Bot. France*, 183–199.

DUBOIS-TYLSKI T. (1973a) Aspects ultrastructuraux de l'induction sexuelle chez *Closterium moniliferum* (Bory) Ehr. *Beih. Nova Hedwigia*, **42**, 91–101.

DUBOIS-TYLSKI T. (1973b) La conjugation en culture 'in vitro' chez *Closterium rostratum* Ehr. *Bull. Soc. Bot. Fr.* **120**, 33–41.

DUBOIS-TYLSKI T. (1978) Evolution ultrastructurale de la zygospore du *Closterium moniliferum* (Bory) Ehr. au cours de sa germination. *Rev. Algol., N.S.* **13**, 211–224.

DUBOIS-TYLSKI T & LACOSTE L. (1970) Action de la temperature et de l'eclairement sur la reproduction sexual d'un *Closterium* du group *moniliferum*. *C. R. acad. sci Paris*, **270**, 302–305.

DUCELLIER R. (1915) Contribution à l'étude du polymorphisme et des monstruosities chez les Desmidiacées. *Bull. Soc. Bot. Genève*, **7**, 75–118.

DUCELLIER F. (1917) Notes sur la Pyrenoide dans le genre *Cosmarium* Corda. *Bull. Soc. bot. Genève*, **9**, 1–5.

DUTHIE H.C. (1964) The survival of desmids in ice. *Br. Phycol. Bull.* **5**, 376.

DUTHIE H.C. (1965a) A study of the distribution and periodicity of some algae in a bog pool. *J. Ecol.* **53**, 343–351.

DUTHIE H.C. (1965b) Some observations on the ecology of desmids. *J. Ecol.* **53**, 695–703.

EIBL K. (1939) Studien uber Plasmolyscverhalten der Desmidiaceen Chromatophoren. *Protoplasma* **33**, 531–539.

ELFVING F. (1889) Om uppkomsten af taggarue hos *Xanthidium aculeatum* Ehr. *Bot. Notiser.* 208–209.

ETTL, H., MULLER D.G., NEUMANN K., STOSCH H.A. VON & WEBER W. (1967) Vegetative Fortpflauzung, Parthenogenese und Apogamie bei Algen. In: *Hundbuch der Pflanzenphysiologie hosg. von W. Ruhland*, **18**, 597–776. Berlin, Hindelberg and N.Y. Springer.

EVANS J.H. (1958) The survival of freshwater algae during dry periods. Part I An investigation of the algae of five small ponds. *J. Ecol.* **46**, 149–167.

EVANS J.H. (1959) The survival of freshwater algae during dry periods. Part II Drying experiments. Part III Stratification of algae in pond margin litter and mud. *J. Ecol.* **47**, 55–70, 71–81.

FISCHER A. (1883) Ueber die Zell Theilung der *Closterium*. *Bot. Zeit.* **41**, 223–272.

FISCHER A. (1884) Ueber das Vorkommen von Gypskristallen bei den Desmidieen. *Jahrb. f. wiss. Bot.* **14**, 133–184.

FÖRSTER K. (1969) Amazonische Desmidiaceen, I. Teil: Areal Santarern. *Amazoniana*, **2**, 5–232.

FÖRSTER K. (1972) Die Desmidiaceen des Haloplanktons des Valencia-Sees, Venezuela. *Int. Rev. ges. Hydrobiol.* **57**, 409–428.

FÖRSTER K. (1974) Amazonische Desmidieen, 2. Teil: Areal Manes—Abacaxis. *Amazoniana*, **5**, 135–242.

FORSTINGER A. (1978) Die zygotenbildung von *Closterium lunula*. *Nova Hedwigia*, **29**, 179–189.

FOTT, B. (1965) Evolutionary trends among algae and their position in the plant kingdom. *Preslia*, **37**, 117–126.

FOX J.E. (1958) Meiosis in *Closterium*. *Phyc. Soc. Amer. News Bull.* **11**, 63.

FREDERICK V.R. (1977) The environmental significance of the algal floras from three Central Ohio sediment profiles. Ph.D. Thesis. Ohio State Univ. Published by Great Lakes Lab. State U. Coll. Buffalo, N.Y.

FREMY P. & MESLIN R. (1948) Un plancton a *Staurastrum paradoxum* Meyen observé en Normandie. *Bull. Soc. Linn. de Normandie*, **5**, 153–164.

R

FREY A. (1926) Études sur les vacuoles à cristaux des Clostères. *Rev. gen. de Bot.* **38**, 273–286.

FRITSCH F.E. (1912) Freshwater algae In National Antarctic Expedition, 1901–1904. *Natural History.* 6-*Zoo. and Bot.* 1–60.

FRITSCH F.E. (1931) Some aspects of the ecology of freshwater algae. *J. Ecol.* **19**, 233–272.

FRITSCH F.E. (1935) *The Structure and Reproduction of the Algae* Vol. I. Cambridge University Press.

FRITSCH F.E. (1953) Comparative studies in a polyphyletic group: The Desmidiaceae. *Proc. Limn. Soc. Lond.* **164**, 258–280.

FRYER F. (1957) The food of some freshwater cyclopoid copepods and its ecological significance. *J. Animal Ecol.* **26**, 263–286.

GAUTHIER-LIÈVRE L. (1958) Desmidiacées asymétriques. Le genre *Allorgeia* gen. nov. *Bull. Soc. Hist. Nat. Afr. Nord.* **49**, 93–101.

GEITLER L. (1930) Uber die kernteiling von *Spirogyra. Arch. Protist.* **71**, 10–18.

GEITLER L. (1935a) Neue Untersuchungen über die Mitose von *Spirogyra. Arch. f. Protist,* **85**, 101–9.

GEITLER L. (1935b) Untersuchungen über den Kernbau von *Spirogyra* mittels Feulgens Neklealfarbung. *Ber. d. Deutsch Bot. Ges.* **53**, 270–276.

GEITLER L. (1958) Sexueller Dimorphisms bei einer Konjugate (*Mougeotia heterogama* n. sp.). *Öst. bot. Zeit.* **105**, 301–332.

GEITLER L. (1965) *Mesotaenium dodekahedron* n. sp. und die Gestalt und Entstelung seiner Zygoten. *Öst. Bot. Zeit.* **112**, 344–358.

GERRATH J.F. (1966) Giant chromosomes in *Triploceras. Brit. Phycol. Bull.* **3**, 154.

GERRATH J.E. (1969) *Penium spinulosum* (Wolle) nov. comb. a taxonomic correction based on cell wall ultrastructure. *Phycologia,* **8**, 109–118.

GERRATH J.F. (1970) Ultrastructure of the connecting strands in *Cosmocladium saxonicum* de Bary (Desmidiaceae) and a discussion of the taxonomy of the genus. *Phycologia,* **9**, 209–215.

GERRATH J.F. (1975) Notes on desmid ultrastructure
  I Cell wall and Zygote wall of *Cylindrocystis brebissonii*
  II The replicate division septum of *Bamubsina brebissonii*
  Beih. Nova Hedwigia, **42**, 103–113.

GIBBS S.P. (1962) The ultrastructure of the chloroplasts of the algae. *J. Ultrastructure Res.* **7**, 418–435.

GIBBS S.P. (1970) The comparative structure of the algal chloroplast. *Ann. N.Y. Acad. Sci.* **175**, 454–473.

GODWARD M.B.E. (1950a) Somatic chromosomes of Conjugales. *Nature,* **165**, 653.

GODWARD M.B.E. (1950b) On the nucleus and nucleolar-organizing chromosomes of *Spirogyra. Ann. Bot.* **53**, 39–54.

GODWARD M.B.E. (1953) Geitler's nucleolar substance in *Spirogyra. Ann. Bot.* **67**, 403–418.

GODWARD M.B.E. (1954) The 'diffuse' centromere or polycentric chromosomes of *Spirogyra. Ann. Bot.* **70**, 143–156.

GODWARD M.B.E. (1956) Cytotaxonomy of Spirogyra I. *S. submargaritata, S. subechinata* and *S. britannica* spp. novae. *J. Linn. Soc. Lond. Bot.* **55**, 532–546.

GODWARD M.B.E. (1966) *The Chromosomes of Algae.* Edw. Arnold, London.

GODWARD M.B.E. & JORDAN E.G. (1965) Electron microscopy of the nucleolus of *Spirogyra britannica* and *S. ellipsospora. J. Roy. Microscop. Soc.* **84**, 347–360.

GODWARD M.B.E. & NEWNHAM R.E. (1965) Cytotaxonomy of Spirogyra II. *S. neglecta, S. punctulatum, S. majuscula, S. ellipsospora* and *S. porticalis. J. Linn. Soc. Lond. Bot.* **59**, 99–110.

GOLDSTEIN S. (ed.) (1952) *Modern developments in fluid mechanics.* Oxford.

GORHAM E. (1956) On the chemical composition of some waters from Moor House nature reserve. *J. Ecol.* **44**, 373–82.

GORHAM E. (1957) The chemical composition of some waters from lowland lakes in Shropshire. *Tellus*, **9**, 174–179.

GOUGH S.B. (1977) The growth of selected desmid taxa at different calcium and pH levels. *Amer. J. Bot.* **64**, 1297–1299.

GOUGH S.B. & WOELKERLING W.J. (1976) Wisconsin Desmids II. Aufuchs and plankton communities of selected soft water lakes, hard water lakes and calcareous spring ponds. *Hydrobiologica*, **49**, 3–25.

GOUHIER M. & TOURTE M. (1969) Nature et role de quelques formations particulières chez *Micrasterias fimbriata*. *J. Micr.* **8**, 50a.

GRIFFITHS B.M. (1928) On desmid plankton. *New Phytol.* **27**, 98–107.

GRIFFITHS D.J. (1970) The pyrenoid. *Bot. Rev.* **36**, 29–58.

GRÖNBLAD R. (1921) New desmids from Finland and N. Russia. *Acta Soc. Fauna Flora Fenn.* **49**, 1–78.

GRÖNBLAD R. (1935) Sub-aerial Desmids II. *Soc. Sci. Fenn. Comm. Biol.* **5**, 1–4.

GRÖNBLAD R. (1945) De algio Brasiliensibus. *Acta. Soc. Sci. Fenn.* **11**, 1–43.

GRÖNBLAD R. (1953) Algological Notes IV. On some desmids and diatoms living in brackish water. *Mem. Soc. Fauna Fl. Fenn.* **28**, 48–50.

GRÖNBLAD R. (1954) *Amscottia* Grönbl. nom. nov. *Bot. Notiser*, **107**, 433.

GRÖNBLAD R. (1956) A contribution to the knowledge of the algae of brackish water in some ponds in the Woods Hole region U.S.A. *Mem. Soc. Fauna Flora Fenn.* **31**, 63–69.

GRÖNBLAD R. (1957) Observations on the conjugation of *Netrium digitus*. *Bot. Notiser*, **110**, 468–472.

GRÖNBLAD R., PROWSE C.A. & SCOTT A.M. (1958) Sudanese desmids. *Acta Bot. Fenn.* **58**, 1–82.

GRÖNBLAD R. & RŮŽIČKA (1959) Zur Systematik der Desmidiaceen. *Bot. Notiser*, **112**, 205–226.

GRÖNBLAD R. & SCOTT A.M. (1955) On the variation of *Staurastrum bibrachiatum* as an example of variability in a desmid species. *Acta. Soc. Fauna Flora Fenn.* **72**, 1–11.

GRUENDLING G.K. (1971) Ecology of the epipelic algal communities in Marion Lake, British Columbia. *J. Phycol.* **7**, 239–249.

HÄDER D.P. & WENDEROTH K. (1977) Role of three basic light reactions in photomovements of desmids. *Planta*, **137**, 207–214.

HAGER A. & MEYER-BERTENRATH T. (1966) Die Isolierung und quantitative Bestimmung der Carotinoide und Chlorophyll von Blattern, Algen und isolierten Chloroplasten mit Hilfe dunnschict-chromatographisher Methoden. *Planta*, **69**, 198–217.

HALFEN L.N. & CASTENHOLZ R.W. (1971) Energy expenditure for gliding motility in a blue-green alga. *J. Phycol.* **7**, 258–260.

HARTMANN M. (1955) Sex problems in Algae, Fungi and Protozoa. *Amer. Nat.* **87**, 321–346.

HAUPT W. (1959) Die Chloroplastendrehung bei *Mougeotia* I. Uber den quantitativen und qualitariven hichtbedarf der Schwachlicht bewegung. *Plant*, **53**, 484–501.

HAUPTFLEISCH P. (1888) Zellmembran und Hüllgallerte der Desmidiaceen. *Mitt. Natur. ver. F. Neuror-pommerr u. Rugen*, **20**, 59–136.

HEILBRON O. (1939) Chromosome studies in the Cyperaceae III-IV. *Hereditas*, **25**, 224–240.

HEIMANS J. (1934) De transportfactor in de sociologie ('accessibiliteit'). *Nederl. Kruidk. Arch.* **44**, 62.

HEIMANS J. (1935) Das Genus *Cosmocladium*. *Pflanzenforschung*, **18**, 1–132.

HEIMANS J. (1942) Triquetrous forms of *Micrasterias*. *Blumea (Suppl.)* **2**, 52–63.

HEIMANS J. (1969) Ecological, phytogeographical and taxonomic problems with desmids. *Vegetatio*, **17**, 50–82.

HERRMANN R.G. (1968) Die Plastiden pigment einiger Desmidiaceen. *Protoplasma*, **66**, 357–368.

HEYWOOD R.B. (1977) A limnological survey of the Ablation Point area, Alexander Island, Antarctica. *Phil. Trans. Roy. Soc., Lond. B.* **279**, 27–38.

HICKEL B. (1973) Limnological Investigations of the Pokhara Valley, Nepa.l *Int. Revue ges. Hydrobiol.* **58**, 659–672.

HILLIARD D.K. (1966) Studies on Chrysophyceae from some ponds and lakes in Alaska I. Notes on the taxonomy and occurrence of phytoplankton in an Alaskan pond. *Hydrobiologia*, **28**, 553–576.

HILTON R.L. (1970) A psychological and morphological study of the algae *Spirotaenia condensata*, Brebisson. Dissertation, University of Arizona, Tucson A.Z., U.S.A.

HIRANO M. (1955-1959) Flora Desmidiarium Japonicum I-VI. *Centr. Biol. Lab. Kyoto Univ.* pp. 1–386.

HIRANO M. (1965) Freshwater algae in the Antarctic. In, van Mieghem J., van Oye, P· and Schull, J. (Eds.) Biography and Ecology in Antarctica. *Monographic Biologicae*, **15**, 127–193.

HOBBIE J.E. (1973) Arctic limnology: a review. In Alaska Arctic tundra. Ed. by M.E. Britton. *Arctic Inst. N. Amer. Tech. Pap.* No. 25, 127–168.

HOFLER K. (1951) Plasmolyse mit Natrium Karbonat. *Protoplasma*, **40**, 426–460.

HOLDEN A.V. (1958) Fertilization experiments in Scottish freshwater lochs II. Sutherland 1954. I. Chemical and botanical observations. *Sci. Invest. Freshwat. Fish. Scot.* No. **24**, 42pp.

HOLMFELD H.R. (1929) Bestrag zur Kenntnis der Desmidiaceen Nordwest deutschlands besonders über Zygoten. *Pflanzenforschung*, **12**, 96 pp.

HORI S. & ITO I. (1959) The annual succession of desmid communities in consquence of organic pollution. *Japan Jour. Ecol.* **9**, 152–154.

HOSHAW R.W. & HILTON R.L. (1966) Observations on the sexual cycle of the saccoderm desmid *Spirotaenia condensata*. *J. Arizona Acad. Sci.* **4**, 88–92.

HOSIAISLUOMA V. (1976) Effect of HCl and NaCl on the growth of *Netrium digitus*. *Ann. Bot. Fennici*, **13**, 170–113.

HUBER-PESTALOZZI G. (1927) Algologische Mitteilung IV. Über Aplanosporen-bilding bei einigen Desmidiaceen. *Arch. Hydrobiol.* **18**, 651–658.

HUGHES-SCHRADER S. & RIS H. (1941) The diffuse spindle attachment of the coccids verified by the mitotic behaviour of induced chromosome fragments. *J. Exp. Zool.* **87**, 429–456.

HUTCHINSON G.E. (1957) *A Treatise on Limnology.* Vol. I. Wiley and Co.

HUTCHINSON G.E. (1967) *A Treatise on Limnology.* Vol. II. Wiley and Co.

HUTCHINSON G.E., PICKFORD G. & SCHUURMANN J.F.M. (1932) A contribution to the hydrobiology of pans and other inland waters of South Africa. *Arch. f. Hydrobiol.* **24**, 1–154.

HYGEN G. (1941) Über eine reversible, in Alkalisalslosungen erfolgende Nukleolen verschenelzung bei *Micrasterias denticulata* Bréb. *Bergens. Mus. Arbok* 1941, 1–34.

HYGEN G. (1943) On plastid and cell-form in *Micrasterias* caused by iron deficiency. *Bergens Mus. Arbok* No. 5, 1–29.

IRÉNÉ-MARIE Fr. (1938) Flore desmidiale de la region de Montréal. Laprairie, Canada, 547 pp.

ICHIMURA T. & WATANABE M. (1974) The *Closterium calosporum* complex from the Rynku Islands—variation and taxonomic problems. *Mem. Nat. Sci. Tokyo*, **7**, 89–102.

JAROSCH R. & KIERMAYER O. (1962) Die Formdifferenzierung von *Micrasterias*-Zellen nach lokaler Lichteinwirkung. *Planta*, **58**, 95–112.

JOHN B. & LEWIS K.R. (1957) Studies in *Periplaneta americana*. *Heredity.* **11**, 1–22.

KAHO H. (1933) Das Verhalten der Pflanzenzellen gegen Schwermetallsalze. *Planta*, **18**, 664–670.

KALLIO P. (1951) The significance of nuclear quality in the genus *Micrasterias*. *Ann. Bot. Soc. Zool.-Bot. Fenn.* '*Vanamo*', **24**, 1–120.

KALLIO P. (1953a) The effect of continued illumination on the desmids. *Arch. Soc.* '*Vanamo*', **8**, 58–72.

KALLIO P. (1953b) On the morphogenetics of the desmids *Bull. Torr. Bot. Club.* **80**, 247–263.

KALLIO P. (1954) A new genus and new species among the desmids. *Bot. Notiser*, **107**, 167–78.

KALLIO P. (1957) Studies on artificially produced diploid forms of some *Micrasterias* species (Desmidiaceae). *Arch. Soc. 'Vanamo'*, **11**, 193–204.

KALLIO P. (1959) The relationship between nuclear quantity and cytoplasmic units in *Micrasterias*. *Ann. Acad. Sci. Fenn.* **44**, 1–44.

KALLIO P. (1961) Mitotic cycles and mitogenesis in *Micrasterias*. *Cytologia*, **26**, 155–169.

KALLIO P. (1963) The effects of ultraviolet radiation and some chemicals on morphogenesis in *Micrasterias*. *Ann. Acad. Sci. Fenn.* **70**, 1–39.

KALLIO P. (1968) On the morphogenetic system of *Micrasterias sol. Ann. Acad. Sci. Fenn.* **123**, 1–23.

KALLIO P. & HEIKKILA H. (1969) UV-induced facies changes in *Micrasterias torreyi*. *Oster. bot. Zeit.* **116**, 226–243.

KALLIO P. & LEHTONEN J. (1973) On the plasmatic template system in *Micrasterias* morphogenesis. *Ann Acad. Sci. fenn.* (*Biol.*) **199**, 1–6.

KARSTEN G. (1918) Über die Tagesperiode der Kern-und Zellteilung. *Zeit. f. Bot.* **10**, 1–20.

KASPRIK W. (1975) Beiträge zur Karyologie der Desmidiaceen—gattung *Micrasterias* Ag. *Beih. Nova Hedwigia*, **42**, 115–137.

KATTNER E., LORCH D. & WEBER A. (1977) Die Bausteine der Zellwand und der Gallerte eines Stammes von *Netrium digitus* (Ehr.) Itzigs. & Rothe. *Mitt. Inst. Allg. Bot. Hamburg*, **15**, 33–39.

KAUFFMANN H. (1914) Über der Eutwicklungsgang von *Cylindrocystis*. *Zeit. f. Bot.* **6**, 721–774.

KELLY M.H., FITZPATRICK L.C. & PEASON W.D. (1978) Phytoplankton dynamics, primary production and community metabolism in a North Central Texas pond. *Hydroboligia*, **58**, 245–260.

KIERMAYER P. (1954) Die Vakuolen der Desmidiaceen ihr Verhabten bei Vitalfärbe -und zentrifugierungsversuchen. *Sitz. Akad. Wiss. Wien Math. Nat. Kl. Abt.* 1, **163**, 175–222.

KIERMAYER O. (1963) Die Rolle des Turgorducks bei der Formbildung von *Micrasterias*. *Ber. d. Deutsch Bot. Gesell.* **75**, 78–81.

KIERMAYER O. (1964) Untersuchungen über die Morphogenese und Zellwandbildung bei *Micrasterias denticulata*. *Protoplasma*, **59**, 76–132.

KIERMAYER O. (1966a) Septumbildung und Cytomorphogenese von *Micrasterias denticulata* nach der Einwirkung von Athanol. *Planta*, **71**, 305–313.

KIERMAYER O. (1967a) Das Septum—Initialmuster von *Micratserias denticulata* und seine Buildung. *Protoplasma*, **64**, 481–484.

KIERMAYER O. (1967b) Dictyosomes in *Micrasterias* and their division. *J. Cell Biol.* **35**, 68A.

KIERMAYER O. (1968a) The distribution of microtubules in differentiating cells of *Micrasterias denticulata*. Bréb. *Planta*, **83**, 223–236.

KIERMAYER O. (1968b) Hemming der Kern—und Chloroplaster migration von *Micrasterias* durch Colchizin. *Die Naturwissensch.* **6**, 299–300.

KIERMAYER O. (1970a) Elektronmikroskopische Untersuchungen zum Problem der cyclomorphogenese von *Micrasterias denticulata*. Bréb. *Protoplasma*, **69**, 97–132.

KIERMAYER O. (1970b) Causal aspects of cytomorphogenesis in *Micrasterias*. *Ann. N.Y. Acad. Sci.* **175**, 696–701.

KIERMAYER O. (1972) Beeinflussung der postmitotischen Kern migration von *Micrasterias denticulata* Bréb. duch das Herbizid Trifluralin. *Protoplasma*, **75**, 421–426.

KIERMAYER O. & DOBBERSTEIN B. (1973) Membrankomplexe dictyosomaler Herkunfr als 'Materizen' für die extraplasmatische Synthese und Orientierung von Microfibrillen. *Protoplasma*, **77**, 437–451.

KIERMAYER O. & DOEDEL B. (1976) Elektronenmikroskopische Untersuchengun zum Problem der Cytomorphogenese von *M. denticulata* II. Eingluss von Vitalzentrifugierung auf Formbildung und Feinstruktur. *Protoplasma*, **87**, 179–190.

KIERMAYER O. & JAROSCH O. (1962) Die Formbildung von *Micrasterias rotata* Ralfs und ihre experimentelle Beeinflussung. *Protoplasma*, **54**, 382–420.

KIERMAYER O. & STAEHELIN L.A. (1972) Feinstruktur von Zellwand und Plasmamembran bei *Micrasterias denticulata* nach Gefrierätzung. *Protoplasma*, **74**, 227–237.

KIES L. (1967) Über Zellteilung und Zygotenbildung bei *Roya obtusa*. *Mitt. Staatsmst. allg. Bot. Hamborg*, **12**, 35–42.

KIES L. (1968) Über die Zygotenbildung bei *Micrasterias papillifera* Bréb. *Flora (B)*, **157**, 301–313.

KIES L. (1970a) Elektronenmikroskopische Untersuchungen über Bildung und Struktur der Zygotenwand bei *Micrasterias papillifera*. I. Das Exospor. *Protoplasma*, **70**, 21–47.

KIES L. (1970b) Elektronmikraskopische Untersuchungen über Bildung und Struktur der Zugotenwand bei *Micrasterias papillifera*. II. Die Struktur von Mesospor und Endospor. *Protoplasma*, **71**, 139–146.

KIES L. (1975) Elekrtonenmikroskopische Untersuchungen über die Konjugation bei *Micrasterias papillifera*. *Beih. Nova Hedwigia*, **42**, 139–154.

KING G.C. (1953a) 'Diffuse centromere' and other cytological observations on two desmids. *Nature*, **171**, 181.

KING G.C. (1953b) Chromosome numbers in the desmids. *Nature*, **172**, 592.

KING G.C. (1959) The nucleoli and related structures in the Desmids. *New Phytol.* **58**, 20–28.

KING G.C. (1960) The cytology of the desmids: the chromosomes. *New Phytol.* **59**, 65–72.

KIRK W.L. & COX E.R. (1975) Observations on polymorphism in the green algae *Cosmarium botrytis* Menegh. (Desmidiaceae). *Phykos*, **14**, 35–45.

KIRK W.L., POSTEK M.T. & COX E.R. (1976) The Desmid Genera *Sphaerozosma*, *Onychonema* and *Teilingia*, an historical appraisal. *J. Phycol.* **12**, 5–9.

KLEBAHN H. (1891) Studien über Zygoten I. Die Keimung von *Closterium* und *Cosmarium*. *Jahrb. Wiss. Bot.* **22**, 415–443.

KLEBS G. (1885) Uber Bewegung und Schleinbildung bei der Desmidiaceen. *Biol. Zentralbl.* **5**, 353–367.

KLEBS G. (1896) *Die Bedingungen der Fortpflanzung bei einigen Algen und Pilzen.* Jena, 544 pp.

KNIEP H. (1928) *Die Sexualität der mierdern Pflanzen.* Jena.

KOBAYASHI S. (1973) Relationship between cell growth and turgor pressure in *Micrasterias*. *Bot. Mag. (Tokyo)*, **86**, 309–313.

KOL E. (1942) The snow and ice algae of Alaska. *Smithsonian Misc. Collect.* **101**(16), 1–36.

KOPETZKY-RECHTPERG, O. (1932) Die Nucleolen in kern der Desmidiaceae. *Beih. Bat. Centralbl.* **49**, 686–702.

KOPETZKY-RECHTPERG O. (1931-1933) Über die Kristalle in den Zellen der Gattung *Closterium* Nitzsch. *Beih. Bot. Centralbl.*, **47**, 291–324.

KOPETZKY-RECHTPERG O. (1954) Beobachtungen an Protoplasma und Chloroplasten der Alge *Netrium digitus* bei Kultur unter Lichtabschluss. *Protoplasma*, **44**, 322–331.

KORDE N.V. (1960) Biostratifikatsiys i tipologiya russkikh sapropelei (Biostratification and Classification of Russian Sapropels). Akademiya Nauk SSSR, Moscow, 219 pp. (English transl. by J.E.S. Bradley, British Library Lending Div., Boston Spa).

KORDE N.V. (1966) Algenreste in Seesedimenten. Zur Entwicklungsgeschichte der See und umliegenden Landschaften. *Ergebn. Limnol.*, **3**, 1–38.

KORN R.W. (1969a) Chloroplast inheritance in *Cosmarium turpinii*. *J. Phycol.* **5**, 332–336.

KORN W. (1969b) Nutrition of *Cosmarium turpinii*. *Physiol. Plant.*, **22**, 1158–1165.

KOSINSKAJA E.K. (1960) Conjugate (II). *Flora Plantanum Cryptogamemum URSS*, **5(i)**, 706 pp.

KOVASK V. (1971) On the ecology of desmids I. Desmids and water pH. *Ecsti NSV Tead. Akad. Toim. Biol.* **20**, 166–177.

KOVASK V. (1973) On the ecology of desmids II. Desmid and the mineral content. *Ecsti. NSV. Tead. Akad. Toim. Biol.* **22**, 334–341.

KREBS I. (1951) Beitrage zur Kenntnis des Desmidiaceen—Protoplasten. I. Osmotische Werte. II. Plastiden Konsistenz *Sitz.—Ber. Akad. Wiss. Wien, math.-nat. Kl. Abt I*, **160**, 579.

KREGER D.R. (1957) X-ray interferences of barium sulphate in fungi and algae. *Nature* **180**, 867–868.

KREGER D.R. & BOERE H. (1969) Some observations on barium sulphate in *Spirogyra*. *Acta bot. Neerl.* **18**, 143–151.

KRIEGER W. (1933) Die Desmidiaceen in: *Rabenhorst's Kryptogamenflora*, Vol. 13.

KRIEGER W. (1939) Die Desmidiaceen Europas mit Berucksichtigung der aussereuropaischen Arten 2. In *Rabenhorsts Kryprogamenflora*. 2 *Aufl.* **13 Abt. 1**, 1–117.

KULLBERG R.G. (1971) Algal distribution in six thermal spring effluents. *Trans. Amer. Microsc. Soc.*, **90**, 412–434.

LACALLI T.C. (1973) Cytokinesis in *Micrasterias rotata*. Problems of directed primary wall deposition. *Protoplasma*, **78**, 433–443.

LACALLI T.C. (1974a) Golgi function and cell wall formation in the Conjugales. *Rev. Canad. Biol.* **33**, 125–133.

LACALLI T.C. (1974b) Composition of the primary wall in *Micrasterias rotata*. *Protoplasma*, **80**, 269–272.

LACALLI T.(1974c) Golgi function and cell wall formation in the Conjugales. *Rev. Canad. Biol.* **33**, 125–133.

LACALLI T.C. (1975a) Morphogenesis in *Micrasterias* I. Tip growth. *J. Embryol. exp. Morph.* **33**, 95–115.

LACALLI T.C. (1975b) Morphogenesis in *Micrasterias* II. Patterns of morphogenesis. *J. Embryol. exp. Morph.* **33**, 117–126.

LACALLI T.C. & ACTON A.B. (1972) An inexpensive laser microbeam. *Trans. Amer. Microsc. Soc.* **91**, 236–238.

LACALLI T.C. & ACTON A.B. (1974) Tip growth in *Micrasterias*. *Science, N Y.* **183**, 665–666.

LACK T.J. (1971) Quantitative studies of the Rivers Thames and Kennet at Reading. *Freshwater Biology*, **1**, 213–224.

LACK T.J. & LUND J.W.G. (1974) Observations and experiments on the phytoplankton of Blelham Tarn, English Lake District. 1. The experimental tubes. *Freshwat. Biol.* **4**, 399–415.

LAGERHEIM G. (1892) Die Schnecflora des Pichincha. Ein Beitrage zier Kenntniss der nivalen Algen und Pilzen. *Ber. deutsch. Bot. Ges.*, **10**, 517–534.

LAPORTE L.J. (1931) Recherches sur la biologie et la systematique des Desmidées. *Encycl. Biol.* **9**, 150 pp.

LEBLOND E. (1928) Formation des vacuoles accessoires chez les *Closterium lunula*. *C.R. Acad. Sci. Paris*, **186**, 1311–1314.

LEEDALE G.F. (1968) The nucleus in *Euglena*. In: *The biology of* Euglena (D.E. Buetow, Ed.). Acad. Press N.Y. and London, pp. 185–242.

LEFÈVRE M. (1934) Sur la division et l'elongation des cellules dans le genre *Closterium*. *C.R. Acad. Sci. Paris*, **198**, 1166–1168.

LEFÈVRE M. (1935) Sur la signification des corpuscules trepidants chez les Desmidiées. *Vol. jubilaire du Tricentenaire du Museum*. Paris.

LEFÈVRE M. (1937) Techniquee des cultures cloriques de Desmidiees. *Aun. Sc. Nat. Bot. Ser.* **10**, 19.

LEFÈVRE M. (1939) Recherches experimentales sur le polymorphisme et la teratologie des Desmidiees. *Encycl. Biolog. Paris*, **19**, 42 pp.

LEFÈVRE M. (1964) Extracellular products of algae: in *Algae and Man* (D.F. Jackson, Ed.), **4**, 337–367.

LEFÈVRE M., JAKOB H. & NISBET M. (1952) Auto et heteroantagonisme chez les Algues d'eau douce dans les collections d'eau naturelles. *Ann. Stat. cent. Hydrobiol appl.* **4**, 5–198.

LEFÈVRE M. & NISBET M. (1948) Sur la secretion par certaines espèces d'algues de substances inhibitrices d'autres espèces d'algues. *C.R. Acad. Sci. Paris*, **226**, 107–108.

LEHTONEN J. (1977) Morphogensis in *Micrasterias torreyi* and *M. thomasiana* studied with UV microbeam irradiation and chemicals. *Annals bot. fenn.* **14**, 165–190.

LEMMERMANN E. (1904) Das Plankton schwedisher Gewasser. *Arkiv. Botan.* **2**, 1–209.

LENZENWEGER R. (1968) Zygotenbildung bei der Zieralge *Micrasterias*. *Mikrokosmos*, **57**, 10–13.

LENZENWEGER R. (1973) Über Konjugation und Zygotenkeimung bei *Micrasterias rotata* (Grev.) Ralfs (Desmidiaceae). *Beih. Nova Hedwigia*, **42**, 155–161.

LEWIS W.M. (1976) Surface/volume ratio implications for phytoplankton morphology. *Science N.Y.* **192**, 885–887.

LEYON H. (1954) The structure of chloroplasts III. A study of pyrenoids. *Expr. Cell Res.* **6**, 497–505.

LHOTSKY O. (1948) The production of clamydospores by *Closterium moniliferum*. *Stud. bot. Čechoslov*, **9**, 155–159.

LHOTSKY O. (1973) The production of chlamydospores in the genus *Closterium* (Desmidiaceae) in nature. *Beih. Nova Hedwigia*, **42**, 163–169.

LIND E.M. & CROASDALE H. (1966) Variation in the desmid *Staurastrum sebaldi* var. *ornatum*. *J. Phycol.* **2**, 111–116.

LING H.U. & TYLER P.A. (1972a) The process and morphology of conjugation in desmids especially the genus *Pleurotaenium*. *Br. Phycol. J.* **7**, 65–79.

LING H.U. & TYLER P.A. (1972b) Zygospore germination in *Pleurotaenium*. *Arch. Protistenk.* **114**, 251–255.

LING H.U. & TYLER P.A. (1974) Interspecific hybridity in the desmid genus *Pleurotaenium*. *J. Phycol.* **10**, 225–230.

LING H.U. & TYLER P.A. (1976) Meiosis, polyploidy and taxonomy of the *Pleurotaenium mamillatum* complex. *Br. Phycol. Bull.* **11**, 315–330.

LIPPERT B.E. (1967) Sexual reproduction in *Closterium moniliferum* and *C. ehrenbergii*. *J. Phycol.* **3**, 182–198.

LIPPERT B.E. (1969) The effect of carbon dioxide on conjugation in *Closterium*. *XI Int. Bot. Congr. Abst.* p. 129.

LOEFLER H. (1964) Volgelzug und Crustaceenverbreibung. *Zool. Anz. suppl.* **27**, 311–316.

LORCH D.W. (in press) Desmids and heavy metals I. Uptake of lead by cultures and isolated cell walls of selected species. *Beihefte Nova Hedwigia*, **56**.

LORCH D.W. (1978) Desmids and heavy metals II. Manganese: uptake and influence on growth and morphogenesis of selected species. *Arch. Hydrobiol.*, **84**, 166–179.

LORCH D. & WEBER A. (1972) Über der Chemie der Zellwand von *Pleurotaenium trabecula* var. *rectum*. *Arch. f. Mikrobiol.* **83**, 129–140.

LOTT J.N.A., HARRIS G.P. & TURNER C.D. (1972) The cell wall of *Cosmarium botrytis*. *J. Phycol*, **8**, 232–236.

LOUB W. (1951) Über die Resistenz verschiedener Algen gegen Vitalfarbstoffe. *Sitz. Ber. Akad. Wiss. Wien, math.-nat. Kl., Abt I*, **160**, 829.

LUND J.G.W. (1942) The marginal algae of certain ponds, with special reference to the bottom deposits. *J. Ecol.* **30**, 245–283.

LUND J.W.G. (1959) Buoyancy in relation to the ecology of freshwater phytoplankton. *Brit. phycol. Bull.* **1**, 1–17.

LUND J.G.W. (1965) The ecology of the freshwater phytoplankton. *Biol. Rev.* **40**, 231–293.

LUND J.G.W. (1971) The seasonal periodicity of three planktonic desmids in Windermere. *Mitl. Internat. Verein. Leimnol.* **19**, 3–25.

LUND J.G.W. (1972) Preliminary observations on the use of large experimental tubes in lakes. *Verh. int. Verein. Theor. angew Limnol.* **18**, 71–77.

LUND J.G.W. (1972) Eutrophication. *Proc. Roy. Soc. (B).* **180**, 371–382.

LUND J.W.G. (1973) Phytoplankton as indicators of change in lakes. *Environment and Change*, **2**, 273–281.

LUND H.M. & LUND J.W.G. (1970) Is Windermere peculiar? *Br. Phycol. J.* **5**, 269–207.

LÜTKEMÜLLER J. (1893) Beobachtungen über die Chlorophyllkörper einiger Desmidiaceen. *Ost. bot. Zeit.* **43**, 5–11: 41–44.

LÜTKEMÜLLER J. (1895) Ueber die Gattung *Spirotaenia* Bréb. *Ost. bot. Zeit.* **45**, 1–6; 51–57; 88–94.

LÜTKEMÜLLER J. (1902) Die Zellmembran der Desmidiaceen. *Beitr. Biolog. Pfl.* **8**, 347–414.

LUTMAN B.F. (1910) Cell structure of *Closterium ehrenbergii* and *C. moniliferum. Bot. Gaz.* **49**, 241–254.

LUTMAN B.F. (1912) Cell and nuclear division in *Closterium. Bot. Gaz.* **51**, 401–430.

LYON T.L. (1969) Scanning electron microscopy: a new approach to the Desmidiaceae. *J. Phycol.* **5**, 380–382.

MAEZAWA R. & UEDA K. (1979) Formation and multiplication of green cells in a double alga, *Micrasterias crux-melitensis* (Ehr.) Hass. *Cytologia,* **44**, 849–859.

MAGUIRE B. (1963) The passive dispersal of small aquatic organisms and their colonization of isolated bodies of water. *Ecol. Monogr.* **33**, 161–185.

MAGUITT M. (1925) Karyokinese chez le *Penium. J. Soc. Bot. Reiss.* **10**, 415.

MARCENKO E. (1966) Uber die Witkungen der Gammestrahlen auf Algen (Desmidiaceen). *Protoplasma,* **62**, 157–183.

MAYER F. (1969) Elektronmikroscopische Untersuchungen der Grunalge *Ankistrodesmus baunii* mit Hilfe der Gefrieratztechnik. *J. Ultrastruct. Res.* **28**, 102–111.

MENKE W. (1962) Structure and chemistry of plastids. *Ann. Rev. Plant Physiol.* **13**, 27–44.

MENKE W. (1966) The structure of chloroplasts. In: *The Biochemistry of Chloroplasts.* Vol. I. (T.W. Goodwin, Ed.). *Academic Press.*

MESSIKOMMER E. (1928) Beitrage zur Kenntnis der Algen flora der Kantous Zurich III. Folge. Die Algenvegetation des Hinwilen und oberhoflerriedes. *Viertel jahresschr. Naturforsch. Ges. Zurich.* **73**, (1–2).

MESSIKOMMER V.E. (1938) Beitrag zur Kenntnis der fossilen und subfossilen Desmidiaceen. *Hedwigia,* **78**, 107–201.

MESSIKOMMER E. (1942) Beitrag zur Kenntnis der Algen flora und Algenvegetation des Hochgebirges um Davos. *Beitr. Globot. Landesaufn. Schweiz.,* **24**, 452 pp.

MEYEN F.J.F. (1828) Beobachtungen über einige niedere Algenformen. *Nov. act. physico-medica Akad. Caesar Leop. Carol. Nat.* **14**, 768.

MEYER R.L. & BROOK A.J. (1969) Freshwater algae from Itasca State Park, Minnesota. *Nova Hedwigia,* **16**, 251–266.

MILOVANOVIC D. (1960) Desmidiaceae in the *Sphagnum* peat-mosses in Serbia. *Bioloske Inst. N.R. Srbye,* **3** (8), 1–22.

MILOVANOVIC D. (1963) Desmidiaceae in the *Sphagnum* peat-mosses at Tara and Ostrozub. *Inst. Biol. Beograd,* **6**, 1–12.

MIX M. (1965) Zur Variationsbreite von *Micrasterias swainei* und *Staurastrum leptocladum* sowie über die Bedeutung von Kulturversuchen für die Taxonomie der Desmidiaceen. *Arch. Microbiol.* **51**, 168–178.

MIX M. (1966) Licht und elektronmikroskopische Untersuchengen an Desmidiaceen XII. Zur Feinstruktur der Zellwände und Mikrofibrillen einiger Desmidiaceen vom *Cosmarium*—Typ. *Arch. Mikrobiol.* **55**, 116–133.

MIX M. (1968) Zur Feinstruktur der Zellwände in der Gattung *Penium* (Desmidiaceen). *Ber. Deutsch. bot. Ges.* **80**, 715–721.

MIX M. (1969) Zur Feinstruktur der Zellwände in der Gattung *Closterium* (Desmidiaceae) unter besonderer Berucksichtigung des Porensystems. *Arch. f. Mikrobiol.* **68**, 306–325.

MIX M. (1972) Die Feinstruktur der Zellwände bei Mesotaeniaceae und Gonatozygaceae mit einen fergleichenden Betrachtung der verschiedenen Wandtypen der Conjugatophyceae und über deren systematischen Went. *Arch. Mikrobiol.* **81**, 197–220.

MIX M. (1975) Die Feinstruktur der Zellwände der Conjugaten und ihre systematische Bedeutung. *Beih. Nova Hedwigia,* **42**, 179–194.

MOLLENHAUER D. (1975a) Introduction, Survey and some considerations on the present status of desmid research. *Beih. Nova Hedwigia,* **42**, 1–25.

MOLLENHAUER D. (1975b) Zur Herkunft, Evolution und Phylogenese der Zygnematales (Chlorophyta). Gedanken in Anschluss an eine Analyse der longitudinalen Symmetrie der vegetativen Zellen. *Beih. Nov. Hedwigia.* **42**, 195–258.

MOSS B. (1972) The influence of environment factors on the distribution of freshwater algae. An experimental study I. Introduction and Ca concentration. *J. Ecol.* **60**, 917–932.

MOSS B. (1973a) The influence of environmental factors on the distribution of freshwater algae. An experimental study II. The role of pH and the carbon dioxide—bicarbonate system. *J. Ecol.* **61**, 157–177.

MOSS B. (1973b) The influence of environmental factors on the distribution of freshwater algae. An experimental study III. Effects of temperature, vitamin requirements and inorganic N compounds on growth. *J. Ecol.* **61**, 179–192.

MOSS B. (1973c) The influence of environmental factors on the distribution of freshwater algae. An experimental study IV. Growth of test species in natural lake waters, and conclusion. *J. Ecol.* **61**, 193–211.

MURRAY J. & PULLAR L. (1910) *Bathymetrical Survey of the Scottish Freshwater Lochs.* Vol. I. Edinburgh, 785pp.

NAKANISHI M. & MONSI M. (1976) Factors that control the species composition of freshwater phytoplankton, with special attention to nutrient concentrations. *Int. Rev. ges. Hydrobiol.* **61**, 439–470.

NAUMANN E. (1927) Zur Kritik des Planktonbegriffe. *Ark. Bot.* **21A**, 1–18.

NEWNHAM R.E. (1962) *M.Sc. Thesis U. of London.*

NEUSCHELER W. (1967a) Bewegung und Orientierungsweise bei *Micrasterias denticulata* un Licht I. Zur Bewegungs und Orientierungsweise. *Z. Pfl. Physiol.* **57**, 46–59.

NEUSCHELER W. (1967b) Bewegung und Orientierungsweise bei *Micrasterias denticulata* im Licht II. Photokinesis und Phototaxis. *Z. Pfl. Physiol.* **57**, 151–171.

NIEUWLAND J.A. (1909) Resting spores of *Cosmarium bioculatum. Amer. Mid. Nat.* **1**, 4–8.

NOGUCHI T. (1976) Phosphatase activity and osmuim reduction in cell organelles of *Micrasterias americana. Protoplasma*, **87**, 163–178.

NORDENSKIÖLD H. (1951) Cyto-taxonomical studies in the genus *Luzula* I. *Heriditas*, **37**, 325–355.

NORDSTEDT O. (1873) Bidrag till kannedomen om sydligare Norges desmidieer. *Lunds Univ. Arsskr.* **9**, 1–51.

NYGAARD G. (1949) Hydrobiological studies on some Danish ponds and lakes, Part II. The quotient hypothesis and some new or little known phytoplankton organisms. *Kgl. Dansk. Vidensk. Selskab.* **7**, 1–293.

NYGAARD G. (1976) Desmids from an Arctic salt lake. *Bot. Tidskr.* **71**, 84–86.

ODUM E.P. (1971) *Fundamentals of Ecology.* 3rd Edn. W.B. Sanduers, Philadelphia.

OLRIC K. (1973) Phytoplankton som Forurenings indikatorer i Soer. *Vatten*, **29**, 290–301.

ONDRACEK K. (1936) Experimentelle Untersuchungen über die Variabilitat einigen Desmidiaceen. *Planta*, **26**, 226–246.

PALAMAR-MORDVINTSEVA G.M. (1973) Variability in banded species of the genus *Closterium* Nitzch. in ontogenesis. *Ukr. Bot. Zh.* **33**, 31–38.

PALAMAR-MORDVINTSEVA G.M. (1976) Main trends of Desmid evolution. *Ukr. bot. Zh.* **33**, 225–231.

PALAMAR-MORDVINTSEVA G.M. (1976a) New genera of Desmidiales. *Ukr. bot. Zh.* **33**, 396–398.

PALAMAR-MORDVINTSEVA G.M. (1976b) A taxonomic analysis of the genus *Staurastrum* Meyen. *Ukr. bot. Zh.* **33**, 31–38.

PALAMAR-MORDVINTSEVA (1977) New and rare for the Ukranian SSR representatives of the *Euastrum* genus. *Ukr. bot. Zh.* **34**, 583–587.

PALAMAR-MORDVINTSEVA G.M. & KHISORIEV K. (1979) Desmidiales in the purifying system of Dushanke. *Ukr. bot. Zh.*, **36**, 26–31.

PALAMAR-MORDVINTSEVA H.M. (1978a) Analysis of Desmidiales-flora of the Ukranian Carpathians. *Ukr. Bot. Zh.* **35**, 29–38.

PALAMAR-MORDVINTSEVA H.M. (1978b) Distribution of Desmidiales in bogs of the Ukranian Carpathians. *Ukr. bot. Zh.* **35**, 135–141.

PALAMAR-MORDVINTSEVA H.M. & BURLAKINA N.P. (1973) Variability of some characters in *Cosmarium subtumidum* Nordst. in a culture. *Ukr. Bot. Zh.* **30**, 489–496.

PALMER M.C. (1969) A composite rating of algae tolerating organic pollution. *J. Phycol.* **5**, 78–82.

PARSONS W.M., SCHLICHTING H.E. & STEWART K.W. (1966) In-flight transport of algae and protozoa by selected Odonata. *Trans. Amer. Microsc. Soc.* **85**, 520–527.

PEARSALL W.H. (1921) The development of vegetation in the English lakes considered in relation to the general evolution of glacial lakes and rock, basins. *Proc. Roy. Soc. B*, **92**.

PEARSALL W.H. (1924) Phytoplankton and environment in the English lake district. *Rev. Algol.* **1**, 54–67.

PEARSALL W.H. (1930) Phytoplankton of the English Lakes I. The proportions in the water of some dissolved substance of biological improtance. *J. Ecol.* **18**, 306–320.

PEARSALL W.H. (1932) Phytoplankton in the English Lakes II. The composition of the phytoplankton in relation to dissolved substances. *J. Ecol.* **20**, 241–263.

PÉTERFI L.S. (1972) Variability of Staurastra in natural populations with remarks on its taxonomic and nomenclatural implications. *Rev. Romm. Biol.-Bot.* **17**, 19–28.

PÉTERFI L.W. (1974) Structure and pattern of desmid communities in some Rumanian ombrophilous peat bogs. *Nova Hedwigia*, **25**, 651–664.

PFIESTER O.P. (1976) *Oocardium stratum* a rare (?) desmid (from an Okalhoma stream). *J. Phycol.* **12**, 134.

PICKETT-HEAPS J.D. (1973a) Stereo-scanning electron microscopy of desmids. *J. Microsc.* **99**, 109–116.

PICKETT-HEAPS J.D. (1973b) Cell division in *Cosmarium botrytis*. *J. Phycol.* **8**, 343–360.

PICKETT-HEAPS J.D. (1974) Scanning microscopy of some cultured desmids. *Trans. Amer. Microsc. Soc.* **93**, 1–23.

PICKETT-HEAPS J.D. (1975) *The Green Algae*. Sinauer Assoc., Sunderland, Mass.

PICKETT-HEAPS J.D. & FOWKE L.C. (1970) Mitosis, cytokinesis and cell elongation in the desmid, *Closterium littorale*. *J. Phycol.* **6**, 189–215.

PICKETT-HEAPS J.D. & FOWKE L.C. (1971) Conjugation in the desmid *Closterium littorale*. *J. Phycol.* **7**, 37–50.

PICKETT-HEAPS J.D. & MARCHANT H.J. (1973) The phylogeny of the green algae: A new proposal. *Cytobios*, **6**, 255–264.

PIRSON A. (1961) Sur l'application de la synchronisation à la physiologie cellulaire des algues verts. In *Chemie et physico-chemie des principes immédiats tires des algues*. Coll. intenat C.N.R.S. (Dinard (1960), **103**, 103–118.

PLAYFAIR G.J. (1910) Polymorphism and life-history in the Desmidiaceae. *Proc. Linn. Soc. N.S. Wales*, **35**, 459–495.

PORTER K.R. (1972) A method for the 'in situ' study of zooplankton grazing effects on algal species composition and standing crop. *Limnol. Oceanogr*, **17**, 913–917.

PORTER K.R. (1973) Selective grazing and differential digestion of algae by zooplankton. *Nature*, **244**, 179–180.

POTHOFF H. (1927) Beitrage zur Kenntnis de Conjugaten I. Untersuchungen über die Desmidiaceae *Hyalotheca dissiliens* Bréb, f *minor*. *Planta*, **4**, 261–283.

POTHOFF H. (1928) Zur Phylogenie und Eutuicklungsgeschichte der Conjugaten. *Ber. Deutsch. Bot. Ges.* **46**, 667–673.

PRESCOTT G.W. (1948) Desmids. *Bot. Rev.* **14**, 644–676.

PRESCOTT G.W. (1966) Algae of the Panama Canal and its tributaries. *Phykos*, **5**, 1–49.

PRESCOTT G.W. (1968) *The Algae: a Review*. Houghton Mifflin Co., Boston, U.S.A.

PRESCOTT G.W. & SCOTT A.M. (1952) The Genus *Micrasterias* 2. *Trans. Amer. Microsc. Soc.* **71**, 229–252.

PRINGSHEIM E.G. (1918) Die Kultur der Desmidiaceen. *Ber. Deutsch Bot. Ges.* **36**, 482–485.

PRINGSHEIM E.G. (1930) Die Kultur von *Micrasterias* und *Volvox*. *Arch. f. Protistenk*, **72**, 1–48.

PROCTOR V.W. (1959) Dispersal of freshwater algae by migratory water birds. *Science, N.Y.* **130**, 623–624.

PROCTOR V.W. (1966) Dispersal of Desmids by Waterbirds. *Phycologia*, **5**, 227–232.

PUMALY A. DE (1923) Nouveau mode de division cellulaire chez les Conjuguees, unicellulaires (Desmidiacees) sensu latu. *Comp. R. Acad. Sci. Paris*, **176**, 186–188.

RACIBOSKI M. (1889) Desmidyje Nowe. *Fizyjogr. Akad. Umiej. Krakowie*, **19**, 1–38.

RADCHENKO M.I. (1977) Chemotaxonomic study of pigments in *Chlamydomonas* spp. II. Qualitative composition and quantitive content of pigments under external conditions of the medium. *Ukr. Bot. Zh.* **34**, 596–603.

RALFS J. (1848) *The British Desmidieae*. London, 226 pp.

RAVEN J.A. (1968) Exogenous inorganic carbon sources in plant photosynthesis. *Biol. Rev.* **45**, 167–189.

RAWSON D.S. (1956) Algal indicators of trophic lake types. *Limnol. oceanogr.* **1**, 18–25.

REIF C.B. (1939) The effect of stream conditions on lake plankton. *Trans. Amer. Microsc. Soc.* **58**, 398–403.

REYNOLDS C.S. (1973) The phytoplankton of Crose Mere, Shropshire. *Br. Phycol. J.* **8**, 153–162.

REYNOLDS C.S. & BUTTERWICK C. (1979) Algal Bioassay of unfertilized and artificially fertilized lake water maintained in Lund Tubes. *Arch. Hydrobiol. Suppl.* **56**, (Algol. Studies 23), 166–183.

REYNOLDS N. (1940) Seasonal variations in *Staurastrum paradoxum*. *New Phytol.* **39**, 86–89.

RIETH A. (1970) Süsswasser—algenarten in Einzeldarstellung II. *Oocardium stratum* nach Material aus Kuba. *Kulturpflanze*, **18**, 51–71.

ROBINSON D.G., GRIMM I. & SACHS H. (1976) Colchicine and microfibril orientation. *Protoplasma*, **89**, 375–380.

RODHE W. (1948) Environmental requirements of freshwater algae. Experimental studies in the ecology of phytoplankton. *Symbol. Bot. Upsal.* **10(1)**, 1–149.

ROSE F. (1953) A survey of the ecology of the British lowland bogs. *Proc. Linn. Soc. Lond.* **164**, 186–211.

ROSENBERG M. (1940) Formation and division of binucleate giant cells in *Micrasterias americana* (Ehr.) Ralfs. *New Phytol.* **39**, 80–85.

ROSENBERG M. (1944) On the variability of the desmid *Xanthidium subhastiferum* West. *New Phytol.* **43**, 15–22.

ROUND F.E. (1953) The benthic algae communities in Malham Tarn. *J. Ecol.* **41**, 174–197.

ROUND F.E. (1957) Studies on bottom living algae in some lakes of the English lake district III. Distribution on the sediments of algal groups other than Bacillariophyceae. *J. Ecol.* **45**, 649–664.

ROUND F.E. (1963) The Taxonomy of the Chlorophyta. *Br. phyc. Bull.* **2**, 224–235.

ROUND F.E. (1971) The Taxonomy of the Chlorophyta II. *Brit. Phycol. J.* **6**, 235–264.

ROUND F.E. & BROOK A.J. (1959) The phytoplankton of some Irish loughs and an assessment of their trophic status. *Proc. Roy. Irish Acad.* **60B**, 167–191.

RUTTNER F. (1930) Das Plankton des Lunzer Untersees seine Verteilung in Raum und Zut wahrend der Jahre 1908-1913. *Int. Revue gest Hydrobiol. Hydrogr.* **23**, 1–138; 161–287.

RUTTNER F. (1953) *Fundamentals of Limnology* (Transl. Frey & Fry). U. of Toronto Press, 242 pp.

RŮŽIČKA J. (1970) Zur Taxonomie und Variabilität der Familie Gonatozygaceae 1-2. *Preslia*, **42**, 1–15.

RŮŽIČKA J. (1973) Die Zieralgen des Naturschutzgebietes 'Rezahinec'. *Preslia*, **45**, 193–241.

RŮŽIČKA J. (1973) Über einige ganz alltagliche Problem der Desmidiologie und der Desmidiologen. *Beih. Nova Hedwigia*, **42**, 259–273.

Růžička J. (1974) Erwägungen über Taxonomie der Gattung *Closterium* I. *Preslia*, **47**, 193–210.

Růžička J. (1976) Erwägungen über die Taxonomie der Gattung *Closterium* II. *Preslia*, **48**, 1–16.

Růžička J. & Pouzar Z. (1978) Erwägungen über die Taxonomie und Nomenklatur der Gattung *Actinotaenium* Teil. *Folia Geobot. Phytotax, Praha*. **13**, 33–66.

Sampaio J. (1949) Desmidias novas para a flora portuguesa. *Boll. Soc. Broteriana*, **23**, 105–117.

Sano M. & Ueda K. (1980) Relationship between nucleo-cytoplasmic ratio and final cell volume in *Micrasterias crux-melitensis* (Desmidaceae, Chlorophyta). *J. Phycol.* **16**, 52–56.

Sasaki K. (1978) Non-histone proteins associated with chromatin of *Spirogyra* during the conjugation process. *Physiol. Pl.* **42**, 257–260.

Sassen A., Van Eyden-Emons, Laers A. & Wanaka F. (1969) Cell wall formation in *Chlorella pyrenoidosa*: A freeze etching study. *Cytobiologie*, **1**, 333–382.

Schlichting H.E. Jr (1961) Viable species of algae and Protozoa in the atmosphere. *Lloydia*, **24**, 81–88.

Schlichting H.E. Jr (1964) Meteorological conditions affecting the dispersal of airborne algae and protozoa. *Lloydia*, **27**, 64–78.

Schmidt G.W. (1970) Numbers of bacteria and algae and their inter-relations in some Amazonian waters. *Amazoniana*, **2**, 393–400.

Schmidt G.W. (1973) Primary production of phytoplankton in three types of Amazonian waters. I. Introduction. *Amazoniana*, **4**, 135–138.

Schmidt G.W. (1976) Primary Production of Phytoplankton in the three types of Amazonian waters. IV. On the primary production of Phytoplankton in the bay of the Lower Rio Negro. *Amazoniana*, **5**, 517–528.

Schmidt G.W. & Uherkovich G. (1973) Zur Artenfülle des Phytoplanktons in Amazonian. *Amazoniana*, **4**, 243–252.

Schröder B. (1900) *Cosmocladium saxonicium* de Bary. *Ber. Deutsch Bot. Ges.* **18**, 15–23.

Schröder B. (1902) Untersuchungen über Gallerbildung der Algen. *Verh. Naturh.-Med. Ver. Heidelberg*, **7**, 139–196.

Schroter K., Lauchli A. & Sievens A. (1975) Mikroanalytische Identifikation von Barium-sulphat-Kristallen in den Statolithen der Rhizoide von *Chara fragilis* Desm. *Planta*, **122**, 213–225.

Schülle H.H. (1970) Qualitative und quantitative Untersuchungen über das Phytoplankton des Titisees in jahres zeitlichen Verlauf sowie einige Bermerkungen zum derzeitigen Zutsand des Sees. *Arch. Hydrobiol./Suppl.* **38**, 170–211.

Schülle H.H. (1975) Untersuchungen zum synchronen Zellwachstsum von *Staurastrum pingue* Teiling (Desmidiaceae) im Licht-Dunkel-Wechsel. *Beih. Nova Hedwigia*, **42**, 275–281.

Schulz P. (1930) Zur Zygosporenbildung zweier Desmidiaceen. *Ber. West pruss. Bot.-Zool. Ver.* **52**, 17–23.

Schumann C. (1875) Ueber die Bewegungen der Kristallausscheidung im Zellsafr der Desmidiaceen. *Planta*, **26**, 67–72.

Scott A.M. & Prescott G.W. (1956) Notes on Indonesian freshwater algae I. *Reinwardtia*, **3**, 351–362.

Scott A.M. & Prescott G.W. (1956) Notes on Indonesian freshwater algae II. *Ichthyadontum* a new desmid genus from Sumatra. *Reinwardtia*, **4**, 105–112.

Scott A.M. & Prescott G.W. (1961) Indonesian desmids. *Hydrobiologia*, **17**, 1–132.

Selman G.G. (1966) Experimental evidence for the nuclear control of differentiation in *Micrasterias*. *J. Embryol. Exp. Morph.* **16**, 469–485.

Selman G.G. (1972) Anucleate development in *Micrasterias* induced using an ultraviolet microbeam. In: *Biology and radiobiology of anucleate systems II. Plant cells.* (S. Bonotto *et al.* Eds), Academic Press. pp. 145–164.

Shapiro J. (1973) Blue green algae: why they become dominant. *Science*, **179**, 382–384.

260   Bibliography

Sheath R.G. & Hellebust J.A. (1978) Comparison of algae in the euplankton, tycho-plankton, and periphyton of a tundra pond. *Canad. J. Bot.* **56**, 1472–1483.

Simola L.K. & Haapala E. (1970) On the effect of some phenothiazine derivatives,. antihistamines and thalidomide on reproduction and morphogenesis in *Micrasterias torreyi* and *M. sol. Ann. Acad. Sci. fenn. A. IV Biol.*, **172**, 1–14.

Soeder C.J., Schulze G. & Thiele D. (1967) Ein fluss verschiedeuer Kulteerbedingung-en auf das Wachstum in Sychronkultierch von *Chlorella fusca. Arch. Hydrobiol. Suppl.* **33**, 127–171.

Soeder C.J., Muller H., Payer H.D. & Schulle H. (1971) Mineral nutrition of plank-tonic algae: some considerations, some experiments. *Mitt. Intern. Verein. Limnol.* **19**, 39–58.

Soeding H., Doeffling K. & Mix M. (1976) Formation of bactericidal substances by *Cosmarium impressulum. Arch. Microbiol.* **108**, 153–157.

Spence D.H.N. (1964) The macrophyte vegetation of lochs, swamps and association fens. In: *The Vegetation of Scotland* (J.H. Burnett, Ed.), Oliver and Boyd, Edin., pp. 306–425.

Spence D.H.N. (1967) Factors controlling the distribution of freshwater macrophytes with particular reference to the lochs of Scotland. *J. Ecol.* **55**, 147–170.

Staehelin L.A. & Kiermayer O. (1973) Membrane differentiation in the Golgi complex of *Micrasterias denticulata* Bréb. visualized by freeze-etching. *J. Cell Sci.* **7**, 787–792.

Stahl E. (1880) Über den Einfluss von Richtung Stanke der Beleuchtung auf einige Bewegungserscheinungen im Pflanzenriche. *Bot. Zeit.* **38**, 297–304.

Starr R.C. (1954) Heterothallism in *Cosmarium botrytis* var. *subtumidum. Am. J. Bot.* **41**, 601–607.

Starr R.C. (1954) Inheritance of mating type and a lethal factor in *Cosmarium botrytis* var. *subtumidum* Wittr. *Proc. Nat. Acad. Sci.* **40**, 1060–1063.

Starr R.C. (1955a) Isolation of sexual strains of placoderm desmids. *Bull. Torrey Bot. Club*, **82**, 261–265.

Starr R.C. (1955b) Zygospore germination in *Cosmarium botrytis* var. *subtumidum. Amer. J. Bot.* **42**, 577–581.

Starr R.C. (1958a) The production and inheritance of the triradiate form in *Cosmarium turpinii. Amer. J. Bot.* **45**, 243–248.

Starr R.C. (1958b) Asexual spores in *Closterium didymotocum. New Phytol.* **57**, 187–190.

Starr R.C. (1959) Sexual reproduction in certain species of *Cosmarium. Arch. Protis-tenk.* **104**, 155–164.

Starr R.C. & Rayburn W.R. (1964) Sexual reproduction in *Mesotaenium kramstai. Phycologia*, **4**, 23–26.

Steinecke Fr. (1926) Die Gypseristalle der Closterien als statolithen. *Bot. Arch.* **14**, 312–318.

Stewart K.D. & Mattox K.R. (1975) Some aspects of mitosis in primitive green algae: phylogeny and function. *BioSystems*, **7**, 310–315.

Stewart K.D. & Mattox K.R. (1978) Structural evolution in the flagellated cells of green algae and land plants. *BioSystems*, **10**, 145–152.

Swale E.M.F. (1968) The phytoplankton of Oak Mere, Cheshire, 1963-66. *Br. Phycol. Bull.* **3(3)**, 441–449.

Symeons J.J. (1951) Esquisse d'un system des associations algales d'eau douce. *Proc. Int. Limnol. Theor. and appl.* **11**, 395–408.

Taft C.E. (1978) Structural features of double zygospores of *Closterium lineatum. Trans. Amer. Microsc. Soc.* **97**, 594–595.

Talling J.F. (1976) Phytoplankton: Composition, Development and Productivity, in the Nile, Biology of an Ancient River. (Ed. J. Rzsóka) Junk, The Hague, pp. 385–402.

Talling J.F. & Talling I.B. (1965) The chemical compostiion of African lake waters. *Int. Rev. ges. Hydrobiol. Hydrogr.* **50**, 421–463.

Tan K. & Ueda K. (1978) Rapid degeneration of the protoplasm in artificially induced small cells of *Micrasterias crux-melitensis. Protoplasma*, **97**, 61–70.

TANAKA N. (1939) Chromosome studies in the Cyperaceae IV. Chromosome numbers in *Carex* species. *Cytologia*, **10**, 51–58.

TASSIGNY M. (1966) Étude critique de genre *Closterium* (Desmidiales), le group *setaceum* *–kuetzingii. Rev. Algol. N.S.* **8**, 228–250.

TASSIGNY M. (1971a) Observations sur les besoins en vitamines des Desmidiées. *J. Phycol.* **7**, 213–215.

TASSIGNY M. (1971b) Action du calcium sur la croissance de Desmidiées axeniques. *Mitt. internat. verein. Limnol.* **19**, 292–313.

TASSIGNY M. (1973) Observations des variations qualitative des populations de Desmidiées dans quelques étangs mesotrophe et dystrophes. *Beih. Nova Hedwigia*, **42**, 283–316.

TASSIGNY M. & LEFÉVRE M. (1971) Auto, hétéroantagonisme et cultures conséquences des excrétions d'algues d'eau douce ou Thermale. *Mitt. internat. verein. Limnol.* **19**, 26–38.

TAYLOR A.O. & BONNER B.A. (1967) Isolation of phytochrome from the alga *Mesotaenium. Plant Physiol.* **42**, 762–766.

TEILING E. (1916) En Kaledonisk fytoplankton-formation. *Svensk. Bot. Tidskr.* **10**, 506–519.

TEILING E. (1947) *Staurastrum planctonicum* and *S. pingue*. A study of planktic evolution. *Svensk. Bot. Tidsk.* **41**, 218–234.

TEILING E. (1948) *Staurodesmus*, genus novum. *Bot. Notiser*, **101**, 49–83.

TEILING E. (1950) Radiation in desmids, its origin and consequences as regards taxonomy and nomenclature. *Bot. Notiser*, **103**, 299–327.

TEILING E. (1952) Evolutionary studies on the shape of the cell, and the chloroplasts in desmids. *Bot. Notisier*, **105**, 264–306.

TEILING E. (1954) *Actinotaenium*, genus Desmidiacearum resuscitatum. *Bot. Notiser*, **100**, 376–425.

TEILING E. (1956) On the variation of *Micrasterias mahabuleshwarensis* f. *wallichii. Bot. Notiser*, **109**, 260–274.

TEILING E. (1957b) Morphological investigations of asymmetry in desmids. *Bot. Notiser*, **111**, 49–82.

TEILING E. (1967) The desmid genus *Staurodesmus*—a taxonomic study. *Ark. Bot. Ser. 2*, **6**, 467–629.

TEWES L.L. (1969) Dimorphism in *Cosmarium botrytis* var. *depressum. J. Phycol.* **5**, 270–271.

THIERY J.P. (1967) Mise en évidence des polysaccharides sur coupes fines et microscopie électronique. *J. Microsc.* **6**, 987.

THOMASSON K. (1971) *Amazonian algae*. Institut. Roy. des Sci. Naturell. de Belgique Memoires (Deuxieme Ser. Fasc. 86).

THOMASSON K. (1973) Notes on the plankton of some Sydney reservoirs, with descriptions of two interesting desmids. *Contrib. N.S.W. Natl. Herb.* **4**, 384–394.

THOMASSON K. (1973) *Actinotaenium, Cosmarium* and *Staurodesmus* in the plankton of Roturua lakes. *Svensk. Bot. Tidskr.* **67**, 127–141.

THUNMARK S. (1945) Zur Soziologie des Süsswasserplanktons. Eine methodologischokologische Studie. *Fol. Limnol. Scandin.* **3**, 66 pp.

TIPPIT D.H. & PICKETT-HEAPS J.D. (1974) Experimental investigations into morphogenesis in *Micrasterias. Protoplasma*, **81**, 271–296.

TOMASZEWICZ & GRAZYNA H. (1973) The typical variety and developmental stages of *Micrasterias truncata. Acta. Soc. Bot. Pol.* **42**, 591–598.

TOMASZEWICZ G.H. (1974) Desmids of a dune-surrounded lake in Zieleniec near Warsaw. *Acta. Soc. Bot. Pol.* **43**, 399–419.

TOURTE M. (1971) Indentification et role des vesicles de type 'peroxysome' chez *Micrasterias fimbriata. J. Microsc.* **11**, 96.

TOURTE M. (1972) Mise en evidence d'une activite catalasique dans les peroxysomes de *Micrasterias fimbriata. Planta*, **105**, 50–59.

TOURTE M. (1972a) Modifications morphogénétiques indicites par la puromycine et la cycloheximide sur le *Micrasterias fimbriata* Ralfs, au cours de bourgeonnement. *C.R. Acad. Sci. Paris*, **274**, 2295–2298.

TOURTE M. (1972b) Mise en evidence d'une activite catalasique dans les peroxysomes de *Micrasterias fimbriata. Planta*, **105**, 50–59.

TRÖNDLE A. (1907) Ueber die Kopulation und Keimung von *Spirogyra. Bot. Zeit.* **65**, 187–217.

TURNER C. (1922) The life-history of *Staurastrum dickei* var. *parallelum. Proc. Limn. Soc., London*, **134 Sets**, 59–63.

TYLER P.A. (1971) A simple and rapid technique for surveying size and shape variation in desmids and diatoms. *Br. Phycol. J.* **6**, 231–233.

UEDA K. (1972) Electron microscopical observations on nuclear division in *Micrasterias americana. Bot. Mag. (Tokyo)*, **85**, 263–271.

UEDA K. & NOGUCHI T. (1976) Transformation of the Golgi apparatus in the cell cycle of a green alga *Micrasterias americana. Protoplasma*, **87**, 145–162.

UEDA K. & YOSHIOKA S. (1976) Cell wall development of *Micrasterias crux-melitensis*, especially in hypertonic solutions. *J. Cell. Sci.* **21**, 617–631.

UHERKOVICH G. (1976) Algen aus den Flussen Rio Negro und Rio Tapajos. *Amazoniana*, **5**, 465–515.

URL W. (1955) Resistenz von Desmidiaceen gegen Schwermetallsaka. *Sitz. d. Oest. Akad. Wiss Math. Nat. Kl. Abt.* **1(134)**, 207–230.

VAN DER BEN C. (1970) Ecophysiologie de quelques Desmidiées I. Un mode d'utilisation des nitrates. *Hydrobiologia*, **36**, 419–442.

VAN DER BEN C. (1971) Ecophysiologie des quelques Desmidiées II. Perturbations experimentales de la nitrate-respiration. *Hydrobiologia*, **38**, 165–184.

VAN OYE O. (1926) Le potamoplankton du Ruki au Congo Belge et des pays chaud en général. *Int. Revue ges. Hydrobiol. Hydrogr.* **16**, 1–15.

VIAN B. & ROLAND J.C. (1972) Differenciation des cytomembranes et renouvellement du plasmalemme dans les phenomenes de secretion végétales. *J. Microsc.* **13**, 119–136.

VIDYAVATI Y. & NIZAM J. (1972) Morphological variations under varied cultural conditions, with particular reference to *Euastrum spinulosum* var. *duplo-minor* West & West. *Phykos*, **11**, 10–16.

VIDYAVATI Y. & NIZAM J. (1974) The frequency of binucleate cells in *Euastrum spinulosum* after gamma-rays and X-ray radiation. *Phykos*, **13**, 70–75.

VIDYAVATI Y. & NIZAM J. (1976) Experimental studies on certain desmids I. Investigation of media in desmid culture. *Phykos*, **15**, 95–99.

VILLERET S. (1951a) Le déphasage des divisions cellulaires et de la morphogénèse chez les Desmidiées unicellulaire. *Bull. Soc. Sci. Bretagne*, **26**, 49–58.

VILLERET S. (1951b) Essai sur la sociologie des algues d'eau douce. Methode et technique. *Bull. Soc. Sci. Bretagne*, **26**, 59–72.

VILLERET S. (1961) Les problems de la sociologie des algues d'eau donce. 86 *Congs. Soc. Savantes*, 483–487.

VILLERET S. & SAVOURE B. (1957) Recherches experimentales sur l'autoécologie de quelques Desmidiées acidophiles. *C.R. S. Sceane Sci.* **295**, 34–39.

WAGER D.B. & SCHUMACHER G.J. (1970) Phytoplankton of the Susquehana River near Binghampton, N.Y.—seasonal variations and effect of sewage effluents. *J. Phycol.* **6**, 110–117.

WAHL H.A. (1940) Chromosome numbers and meiosis in the genus *Carex. Amer. J. Bot.* **27**, 458–470.

WALLNER J. (1933) *Oocardium stratum* Naeg. eine wichtigs tuff bildende alge sudbayers. *Planta*, **20**, 287–293.

WALLNER J. (1936) Zur Kenntnis der Gattung *Oocardium. Hedwigia*, **75**, 130–136.

WARIS H. (1950a) Cytophysiological studies on *Micrasterias* I. Nuclear and cell division. *Physiol. Planatarum*, **3**, 1–16.

WARIS H. (1950b) Cytophysiological studies on *Micrasterias* II. The cytoplasmic framework and its mutation. *Physiol. Plantarum*, **3**, 236–245.

WARIS H. (1951) Cytophysiological studies on *Micrasterias* III. Factors influencing the development of enucleate cells. *Physiol. Plantarum*, **4**, 387–409.

WARIS H. (1953) The significance for algae of chelating substances in nutrient solution. *Physiol. Plant.* **6**, 538–543.

WARIS H. (1958) Splitting of the nucleus by centrifuging in *Micrasterias. Suomal. Tiedeskat.* **40**, 1–20.

WARIS H. (1958) Splitting of the nucleus by centrifuging in *Micrasterias. Ann. Acad. Sci. Fenn. A., Biol. IV,* **40**, 1–20.

WARIS H. & KALLIO P. (1957) Morphogenetic effects of chemical agents and microcytoplasmic relations in *Micrasterias. Ann. Acad. Sci. fenn A.* **37**, 1–20.

WARIS H. & KALLIO P. (1964) Morphogenesis in *Micrasterias. Advanc. Morphogenes,* **4**, 45–80.

WARIS H. & ROUHIAINEN I. (1970) Permanent and temporary morphological changes in *Micrasterias* induced by gamma rays. *Ann. Acad. Sci. fenn A. IV Biol.* **167**, 1–13.

WARTENBERG A. (1965) Das Pyrenoid als Ort der Chlorophyllbildung in Algenchloroplasten. *Arch. Mikrobiol.* **52**, 83–90.

WARTENBERG A. & DORSCHEID T.R. (1964) Die helicoidale Struktur des Chloroplasten von *Mesotaenium violascens. Arch. Microbiol.* **49**, 291–304.

WAWRIK R. (1975) Observations in *Closterium rostratum* Ehr., *Pocillomonas flos-aquae* and *Chlorogonium leiostracum* in ponds of the 'Niederostreichische Waldviertel'. *Arch. Protistenk.* **117**, 33–37.

WEBER A. (1969) Über die Chloroplastenfarbstoffe einiger Conjugaten. *Flora*, **160**, 457–473.

WEBER C.I. (1965) Pyrenoid formation induced in *Closterium moniliferum* by centrifugation. *Exp. Cell Res.* **38**, 507–510.

WEBER H.A. (1918) Über spat—und postglaciale lakustrine und fluviatile Ablagerungen in der Wyhraniederung bei Lobstädt und Borna. *Abhandl. Naturwiss. Vereins., Bremen,* **29**, 1–49.

WEHRLE E. (1953) Die Schmuckalge *Cosmarium subquadratum* und deren kaum bekanuter Formenkries. *Beitr. naturkoll. Forsch. Sudwest deutschl.,* **12**, 90–115.

WESENBERG-LUND (1905) A comparative study of the lakes of Scotland and Denmark. *Proc. roy. Soc. Edinb.,* **25**, 401–448.

WEST G.S. (1904) *A Treatise on the British Freshwater Algae.* Cambridge University Press.

WEST G.S. & CARTER N. (1923) *The British Desmidiaceae.* Vol. I. Ray Society, London.

WEST G.S. & FRITSCH F.E. (1932) *British Freshwater Algae.* Cambridge University Press, 534 pp.

WEST W. & WEST G.S. (1898) Observations on the Conjugatae. *Ann. Bot.,* **12**, 29–58.

WEST W. & WEST G.S. (1904) *The British Desmidiaceae.* Vol. I. Ray Society, London.

WEST W. & WEST G.S. (1909) The British freshwater phytoplankton with special reference to the desmid plankton and the distribution of British desmids. *Proc. Roy. Soc. (B),* **81**, 165–206.

WEST W. & WEST G.S. (1912) *The British Desmidiaceae.* Vol. IV. Ray Society, London.

WILBOIS-COLEMAN A.D. (1961) The role of modern culture methods in the study of algal life cycles. *Quart. Rev. Biol.* **36,** 247–253.

WINTER P.A. & BIEBEL P. (1967) Conjugation in a heterothallic *Staurastrum. Proc. Pa Acad. Sci.* **42**, 76–79.

WISSELINGH C. VAN (1911) On the structure of the nucleus and karyokinesis in *Closterium ehrenbergii. Proc. Acad. Sci. Amsterdam,* **13**, 365–375.

WISSELINGH C. VAN (1912) Über die Kernstructur und Kernteilung bei *Closterium. Beih. Bot. Central,* **29**, 409–432.

WOELKERLING W.J. (1976) Wisconsin Desmids I Aufwuchs and plankton communities of selected acid bogs. *Hydrobiologia,* **48**, 209–232.

S

WOELKERLING W.J. & GOUGH S.B. (1976) Wisconsin Desmids III. Desmid community composition and distribution in relation to lake type and water chemistry. *Hydrobiologia*, **51**, 3–32.

WURTZ A. (1942) Observations sur les vacuoles de *Closterium dianae* Ehr. *Rev. Algol.* **13**, 17–28.

YEH P.Z. & GIBOR A. (1970) Growth patterns and motility of *Spirogyra* spp. and *Closterium acerosum. J. Phycol.* **6**, 44–48.

YOUNG W.H. & CAIRNS J. (1971) Microhabitat pH differences from those of surrounding waters. *Hydrobiologia*, **38**, 453–461.

YOUNGMAN R.E., JOHNSON D. & FARLEY M.R. (1976) Factors influencing phytoplankton growth and succession in Farmoor Reservoir. *Freshwater Biol.* **6**, 253–263.

ZHUPANENKO R.P. (1973) Desmidiaceae of the Siversk Donets river water bodies floodplain and the Pechenegian Reservoir. *Urk. Bot. Zh.* **30**, 523–53.

INDEX

Numbers in bold type refer to pages on which illustrations occur.

265